Praise for the previous edition of this book:

"Many opponents of strong environmental legislation... argue that environmental protection prevents the development of energy supplies and destroys jobs. Publications [like this one] help us refute such arguments."

• Morris Udall

"...almost certainly the best single source on the subject in the English language."

• *Alternatives* magazine

"An important step toward framing questions and recommendations for lower cost energy sources for the years ahead."

• Barbara Williams, Executive Director
Congressional Black Caucus

"The direction we should be headed in if full employment really matters to us is the same direction we should be headed if we really care about environmental quality: toward small, decentralized, appropriate technology, renewable energy sources. This has been intuitively obvious to a growing number of people, I believe, but intuition seldom wins arguments. [This work] is an argument-winner."

• Friends of the Earth

Energy, Jobs and the Economy

Energy, Jobs

and the

Economy

Richard Grossman and Gail Daneker
introduction by Harvey Wasserman

Alyson Publications, Inc. • Boston, Massachusetts

Much of this book originally appeared under the title *Jobs and Energy,* published by Environmentalists for Full Employment.

First Alyson Publications paperback edition, 1979

9 8 7 6 5 4 3 2

Library of Congress
Catalog Card No.: 78-75084

Grossman, Richard & Daneker, Gail
Energy, Jobs and the Economy.
Boston, Mass.: Alyson Publications, Inc.
7903 781214

ISBN 0 932870 00 7

Contents

introduction

Two Movements, One Goal

What is good for the environment can also be good for employment. . . . The industry-sponsored divisions between the labor and environmental movements are phony.

Short-run survival in this society means having a job. Where income is not guaranteed, unemployment means families don't eat.

For corporate executives, that translates into political leverage. As long as working people are insecure, those doing the hiring can manipulate them.

That simple reality has been used to divide the labor and environmental movements for years. Now, through the efforts of many dedicated people, that gap is narrowing.

But it is a tough task. For nowhere has corporate manipulation of jobs and income security been more brutally evident than in relation to environmental issues, particularly in the field of energy.

The world energy monopolies are now demanding an expansion program unprecedented in ecological destruction and expense. And, they're demanding stiff increases in energy prices. Oil drilling and refineries, strip mining, coal gasification and generators, liquified natural gas, uranium mining and milling, and the construction of huge atomic reactors — they're all being sold as "essential" to the growth of the industrial economy. And for the production of jobs.

The sales pitch goes two ways.

First the corporations say that building these giant projects will employ thousands at the sites.

Second, and perhaps most important in the long run, Exxon and the others have equated an endless spiral of energy consumption with economic growth. Our industrial society, they say, has been built on cheap energy, much of it electric. Only more power production can promote the spread of industry and ease the employment crisis, they say.

Backed by corporate clout, the equation of energy production with economic growth has become a public relations axiom. It has also become a machine by which corporations divide and conquer their opposition.

Energy, Jobs and the Economy challenges both the axiom and the vehicle.

First and foremost, it demonstrates that the production and consumption of more central-generated energy may shrink rather than expand the jobs market. In the past, it tells us, expanding the power base may have fed the expansion of the industrial machine. But in recent times, large quantities of energy — especially electricity — have served to replace human labor, which in increasingly difficult economic times has had fewer and fewer places to go.

Why?

Because corporations use energy to promote automation, and, in turn, to make human beings "uneconomical". Electricity in and of itself is merely a tool — but in this case the multi-nationals have used it to undercut wages and eliminate human livelihoods. And in a society where people without jobs have nowhere to go, that means human hardship.

Automation, of course, involves expensive initial investments. But capital is not usually the main problem of the multi-nationals. And their major investments have a way of being underwritten by our tax dollars.

Meanwhile, the new equipment doesn't go on strike, doesn't demand raises or cost-of-living increases, and isn't prone to fighting unsafe working conditions.

Nor do the corporations pay the real cost of the electricity that makes it all possible. Through preferential rates, big users — heavy industry — pay less for their power than small users — home owners and tenants. Utilities are also using a new method of financing called Construction Work in Progress, which allows them to charge consumers for mammoth power plants even before they are built. Without such subsidies from the small consumer, huge quantities of central-generated power might well be impossible to come by.

Indeed, the multi-nationals get us coming and going. They use the short-term jobs on power project construction sites as bait to get us to build machines designed to put us out of work. Then they fire up the machines with electricity for which they pay preferential rates, often below cost, at the expense of the very people whose jobs are most threatened.

And yet, that has been the only game in town. Because corporate executives have been calling the shots, working people have been forced to identify with their interests. In the name of a vague goal of jobs and economic growth — without specifying what kinds of jobs, what kinds of economic growth, and *for whom* — the multi-nationals have been able to enlist major unions on the side of unbridled energy expansion.

But the damage doesn't end with higher electric rates and unemployment. Recent decades of head-long growth have left the natural environment in chaos. Though the planetary ecology has long been portrayed as an ephemeral issue, reserved for idealists, we are now learning that its ills translate not only into human disease and death but also into inflated food costs, overwhelming medical bills, unemployment, and, ultimately, the gigantic bill we'll all have to foot for patching it up.

Experience tells us that the executives and stockholders who profited from making the mess won't be the ones paying to clean it up. Working people — the small consumers and common taxpayers — will have to cough up yet again.

How can we break the cycle? How can we guarantee both jobs and a decent planet on which to live?

Energy, Jobs and the Economy provides some important answers, both politically and in terms of economic theory.

In recent years, a highly visible movement has grown up to deal with issues of the natural environment. It is in many ways the natural inheritor of traditions established by the anti-war, civil rights, labor, and farm-populist movements that have come before it.

As a single-issue campaign, however, environmentalism lacked a crucial dimension — it often did not take into account the employment and equity implications of what it proposed.

This led to hard feelings in the labor movement, where jobs

are the leading issue. Ecologists accused labor of putting employment before air and water; unionists called environmentalists snobs who didn't care about working people. Industry loved it.

For neither party, it seems, realized that its interests were inseparable from the other's. In 1975, some people hoping to blow the whistle on this crazy fight formed a group called Environmentalists for Full Employment (EFFE), aimed at identifying the common ground. Among others, Gail Daneker and Richard Grossman set to work building bridges between the two movements. Politically, the ecology campaign could never hope to make significant inroads without strong support from organized labor.

At the same time, a vicious and well-funded right-wing campaign was chopping up labor's support in Congress, pushing so-called right-to-work laws in numerous states and fighting worker rights every step of the way. What EFFE and a number of other groups did was effect the beginnings of a handshake. Environmentalist support was organized for the Labor Law Reform Act, the J.P. Stevens boycott, the coal miners' strike, and other crucial union campaigns; and members of the union movement found themselves increasingly in the company of ecologists and their literature.

What both are learning is that what is good for the environment can also be good for employment, especially in the field of energy.

First of all, at precisely the time when our traditional fossil fuel supplies seem most threatened, and when atomic energy is becoming politically and economically non-viable — we also find that neither is a good employment investment, even in the short run.

It takes more than $100,000 in capital to create one job in the petroleum industry. Roughly the same is required in the nuclear field. Combined, the two industries consume more dollars per job-created than virtually any other in the American economy.

The reason: money that goes into fossil and nuclear production goes mostly to pay for the expensive, high-technology machinery required by the industry. Oil refineries and generators are mostly automated. The $2.5 billion (by optimistic 1978 estimates) Seabrook Nuclear Power Station is scheduled to employ just 200 steady workers from the time of opening until decommissioning. It doesn't take much sophisticated calculation to realize that $2.5 billion spent almost anywhere else would create more jobs.

To be sure, building reactors does put people to work. The most important nuclear carrot has been the ability to get several thousand construction workers on the job.

But those jobs are outrageously expensive. They cost two, four, even ten times as much to create as do comparable jobs in other fields. They are also highly specialized, limited to a mobile elite, inaccessible to the vast majority of the work force. Yet when a multi-national comes to town with blueprints and bank books, it invariably posits the question in terms of building or not building. Do it our way, they say — take the jobs at the site, or face unemployment.

Thankfully, that's not the real limit of our choice. The money those executives are throwing around is *our* money, and there are other ways to use it, especially when it comes to energy.

One is conservation, recycling and increased efficiency. As *Energy, Jobs and the Economy* amply illustrates, American industry throws a solid 50% of its energy out of the window. This country industrialized with the assumption of infinite energy supplies, and has never bothered to tighten up. Inefficient machinery, mis-matched systems, sealed high-rise office buildings, heavy automobiles, lack of public transportation — it all adds up to an energy economy built on waste.

The cheapest way to produce new energy is to save what is being wasted. And closing the gaps and cleaning up that waste is a task of such magnitude, involving so much human-scale work, that it could have a very significant impact — both short-term and long-term — on our employment picture.

It has also become a necessity of last resort. Neither the political climate nor the money for unrestrained energy expansion exists. Costs have soared and the capital has dried up. The 500 nuclear reactors once bandied about on the drawing boards represent so much of our financial resources that they would suck virtually every other growth sector of the economy bone dry. There are already serious questions about how to pay for even the 250 or so now wending their way through the regulatory and construction hopper.

Realistically, nuclear power is no longer an option. In the long run, neither is oil, coal or gas.

As *Energy, Jobs and the Economy* points out, that leaves us the sun.

Way back in 1952, the Blue Ribbon Paley Commission reported to Harry Truman that there could be 13 million-plus solar heated homes in the U.S. by 1975. The following year, Dwight Eisenhower delivered his "Atoms for Peace" pitch at the United Nations, and the solar vision was lost. But had solar power been allowed to develop, there would be millions of people employed in the field today.

As it stands now, solar advocates are at a disadvantage when it comes to talking about jobs. We don't have a multi-billion-dollar industry, with thousands of employees to point to, in the here and now.

What we do have, however, is an industry approaching take-off. ''Solar energy is not an exotic dream of the future,'' says Senator Charles Percy of Illinois. ''It is workable today.''

Senator Percy is joined in that opinion by a growing legion of energy experts, ecologists and unionists. Solar power is on the brink of becoming a mass reality.

And, among others, major labor organizations are beginning to see its employment potential. ''During the oil crisis we commissioned two studies of where the most secure future for our people would be,'' says Ed Carlough, President of the 175,000-member Sheetmetal Workers' International. ''And they both came up with the same answer: Solar energy.''

Indeed, the Sheet Metal and Air Conditioning National Contractors' Association told Congress, ''We have the skills, shop facilities, manufacturing capability and expertise to embrace this industry now. New tools and processes are not required.''

Similar sentiments have been voiced by Bill Winpisinger, President of the International Association of Machinists.

And by Douglas Fraser, President of the United Auto Workers. ''Today,'' says Fraser, ''when our nation needs both safe energy and decent jobs, solar power can provide both. A fully-developed solar industry, accompanied by a national commitment to high energy efficiency, would be able to meet much of the country's energy requirements.

"Moreover, a thriving, diverse and extensive solar industry would greatly reduce and perhaps eliminate the need for oil imports and provide jobs for millions of Americans. It would move this country closer to its crucial goals of energy independence and full employment."

And that, indeed, is what it's all about.

The coalition between the labor and ecology movements has only begun to take root. Both are large, diverse movements, and each contains segments with interests that are often very complex and far apart. Some unions find themselves on opposite sides of the fence from others on environmental questions; some ecology groups have been unable to agree on certain labor questions as well.

One crucial problem has been a union wariness about the type of jobs involved in the conservation/solar field. Because solar power is "labor-intensive", involving relatively simple skills and high work input, it will offer jobs to the broad unskilled and semi-skilled majority of the work force.

This majority is for the most part unorganized, and many unions fear a major shift to solar might undercut the heavy industries in which they have been traditionally organized.

The answer to that must come from both the labor and environmental camps. Unions must step up their efforts to organize the unorganized, to work harder among the unskilled and semi-skilled, and in the new industries, to guarantee the rights of the people on those jobs. By the same token, pro-solar, pro-conservation activists will have to be equally conscious of work conditions, standards, and levels of organization within the growing new fields. It will not do to establish an environmentally sound energy economy that exploits the people who build it.

Because they are so big and so thoroughly dominated by multi-national corporations, fossil and nuclear generators have been a major barrier to democracy in the workplace, and in the energy economy as a whole.

Our excessive reliance on them has also made this country far too dependent on outside supplies of energy, over which we ultimately have no control.

Solar power, on the other hand, is based on small-scale, decentralized modes of production and installation. Its material inputs are simple, its natural mode of production and installation accessible to the general population, and its fuel is domestically available and in the public domain, immune from the winds of war and inflation. In the long run, solar power is our best — probably our only — hope for a democratic energy economy, owned and operated by the working community as a whole.

The industry-sponsored divisions between the labor and environmental movements are phony. Everyone's right to a clean, cheap, reliable and locally-owned power supply is just as inalienable as each person's right to a safe job and a decent living. When the two movements finally do get totally together on these issues, they should be able to rock this country to its core.

As a society, we have been told time and again that without huge leaps in energy production and consumption, our economy will suffer and unemployment will result.

We have also been told that the destruction of our health

and environment is a price we must pay for economic well-being. What *Energy, Jobs and the Economy* tells us is that both these statements are patently false. Traditional energy expansion destroys both jobs and the planet. Only expansion of solar power and conservation can move us towards a guaranteed energy supply which working people and communities can control, and towards a society in which jobs are guaranteed for all.

The publication of *Energy, Jobs and the Economy* provides an important tool for taking that message — and those goals — out into the political fray.

Harvey Wasserman
Montague, Massachusetts

chapter

1 Jobs and Energy: What's the Connection?

*Energy expansion is **not** necessary for economic and job security. The connection between the two is less simple than the energy corporations would like us to believe.*

Corporate energy interests, along with most industrialists and some government agencies, are vigorously urging the rapid expansion of energy production. The energy systems they want are large in scale, technologically complex, and will be costly, wasteful, environmentally destructive and dangerous to energy industry employees and to the public.

An increasing number of Americans believe that these systems — such as nuclear fission and fusion, conversion of coal and shale to gas and oil, and expanded coal-fueled electric generation — are *too destructive* to the public's health — as well as to workplace and natural environments — to be acceptable. These citizens propose instead a large decrease in the nation's waste of energy, plus immediate utilization of proven solar energy technologies and development of solar technologies almost ready to be utilized.

Most proponents of these large-scale, complex energy systems concede that these systems will expose the public to a variety of dangers. But, they contend, there is absolutely no other way to meet the nation's energy needs, to provide for a strong economy, and to create sufficient numbers of jobs.

We disagree. There *is* another way.

By wasting less energy, plus developing solar sources of energy, our country *can* provide sufficient energy for a prosperous economy. In fact, such a solution to the nation's energy problem *actually leads to a more stable economy and to more jobs* than would the large-scale systems. It does so with less pollution, less disease, less social disruption, and less interference with community, labor union and individual rights.

Decisions on the nation's economic, energy and employment futures are being made now. Wrong decisions today will be irreversible: if the nation decides to pin its hopes on inefficient, large-scale energy systems, these systems will consume so many resources and so much money and cause so much havoc that energy and job options for the future will be choked off.

In sum, a future based on the visions of America's giant energy corporations and some of the federal agencies charged with planning energy strategies will be neither prosperous nor safe. Americans need to know this; and they need to know there is an alternative approach which will provide the nation with safe energy, with prosperity and with jobs.

Energy and the Economy

The latest recession is the sixth since World War II. It is the most severe.[1] Total real unemployment is between 8% and 10%. Women, minorities and young people are out of work in even higher percentages. In 1976, incomes of three million additional Americans fell below the "poverty line." Rural poverty has increased. The nation's largest cities are facing severe financial crises, and have cut back a broad range of vital human services.[2] Industry has been operating at less than full capacity, and inflation has cut deeply into most wage increases.

Americans have long been told that ever-increasing energy production was the key to national economic well-being and jobs. As energy production expanded over the years, so did economic growth and total employment. Many in government and industry are therefore saying that unemployment can be ended only by stepping-up energy development to the greatest degree possible, and with the largest systems possible. Officials throughout the federal government, and especially at the Department of Energy, along with many people in Congress, at industry trade groups like the Edison Electric Institute and the Atomic Industrial Forum, and at the state and local level, are urging such policies.

Yet, current high unemployment, along with a succession of economic crises, have been taking place *while national energy use has been at an all-time high, and increasing.*

The widening gap between accepted theory and reality has prompted calls for re-evaluation of the relationships between energy expansion, economic health and jobs. Cass Alvin, of the United Steelworkers of America, reflected this new outlook recently at a Los Angeles Federation of Labor (AFL-CIO) meeting. The audience had been listening to threats of job loss delivered by California utility executives. Alvin said:

> I think we've got to deal with the nature of energy and jobs very carefully, or we'll find ourselves looking to the wrong allies when confronting energy issues....[3]

And an economist from the International Woodworkers of America (AFL-CIO/CLC) has warned in a similar vein:

> Unions risk disaster and perform a disservice to their members if they do not attempt to define [energy and job relationships.][4]

It sometimes appears deceptively easy to determine which event causes another, especially when there are some apparent connections. For example, some wags have pointed out that over the years there has been a direct correlation between the salaries of a certain denomination of ministers in Massachusetts and the price of rum in Cuba.[5] Yet, no one really believes that a real connection exists between these two occurrences. Similarly, along with energy growth during this century there has been an increase in mental illness. But it is not likely that advocates of energy expansion would say that such tragedy is connected to or necessary for continued energy growth.

To begin probing the nature of energy-employment relationships, we will look into the manner in which energy has been used.

The Substitution of Energy for Labor

Historically, industry has tried to substitute energy for human labor. The amount each working person could produce has therefore increased steadily. But after energy was substituted for labor in a process, fewer workers were needed. The total number of workers could rise only if the demand for products also rose. In other words, the increased demand would have to create more jobs than were eliminated by the energy substitution.

For instance, assume the Lutkus Felt Company discovered it could increase profits by replacing most of its employees with machinery (which involves investment of energy, resources and dollars). Once the machinery is installed, Lutkus can lay off its employees. Rival felt companies need to stay competitive with Lutkus; they also automate and lay off employees.

So after Lutkus and its competitors have finished installing their new machines, the result is:

1. energy consumption in the felt industry is up;
2. employment in the felt industry is down.

Such has been the pattern of American industrial development. What has kept the "more energy leads to more jobs" myth alive has been that accompanying a growing population has been a very large increase in the use of goods and services per person. There has also been a significant increase in energy use. It has thus *appeared* as if energy expansion had been causing economic expansion and increases in jobs. But actually, *constantly expanding demand* has led to constantly expanding production and employment. As Louisiana State University's Professor Herman Daly has concluded:

> Clearly, what is responsible for increasing total employment is the increase in total [goods and services], not the increase in inanimate power production, which by itself must decrease employment.[6]

Productivity and Jobs

From industry's point of view, energy and investment dollars are better than human labor. Automated equipment does not complain about unsafe working conditions, seek wage hikes, question the nature of jobs or products, or strike. And energy has been cheap, subsidized by the government and by the American people.

Goods-producing industries have therefore sought to use fewer workers. They have tried to increase output per worker, or "productivity," and total employment in the manufacturing sector has declined over the last 40 years. Organized labor has supported this emphasis on "productivity," and has allowed the wage scales of skilled labor to be linked to the "productivity index." To management, this index is an important indicator of how well an industry is performing.

But as Hazel Henderson of the Princeton Center for Alternative Futures has pointed out, the "productivity index" is really an "automation index" — a guide to tell management how well it is doing in its drive to get rid of employees. Thus, as more and more energy is substituted for labor, automation increases and jobs decrease (unless there is an accompanying switchover to shorter work days, and a basic redistribution of wealth with increased consumer access to goods and services). The "productivity index" goes up, and management knows it is being successful.

Labor's acceptance of "productivity" — that is, automation — as a criterion for wage increases has caused a widening of the earnings gap between those employed in high energy industries (who are members of well-organized, powerful unions) and those working in areas where energy (and skill) requirements are low (and whose unions are less powerful).[7]

It has also caused the need for more and more energy, and a *decrease* in the total number of industrial jobs. In 1971, the five largest manufacturing industries (primary metals; stone, clay and glass; food; chemicals; and paper) provided only 7.3% of the nation's jobs. From 1950 to 1970, there had been no employment growth in these industries; yet their gross energy consumption during these years increased greatly.[8]

In the steel industry from 1949-1969, the number of production jobs decreased by 20%, while steel input increased by 45%[9]

The aluminum industries in the Pacific Northwest provide a particularly glaring example of high-energy, low job ratios. According to the Bonneville Power Administration, this industry consumes *25%* of the region's electricity, and provides but *one-half of one percent* of the total jobs in that region.[10]

In agriculture, the use of energy for fertilizers, chemicals and automated equipment increased the output per worker. But this led to a steep decline in the number of people employed. In 1970, agricultural employment was only a third of what it had been in 1920. That year, about 27 billion person-hours of labor were needed, compared with only about 2 billion person-hours 50 years later. Energy input increased more than 4 times over that period.[11]

Between 1950 and 1971, total national employment increased 41% while jobs in the energy-producing industries increased only 5.5%.[12] And this small increase consisted mostly of gas station employees. Today, most of these jobs are vanishing, being replaced by customers and self-service gas pumps.

From 1961-1973, electric utilities increased their kilowatt output about 130%, their revenues about 260%, their construction costs about 340%. But *employment* in electric utilities increased only 21%.[13]

Bruce Hannon, at the University of Illinois' Center for Advanced Computation, has calculated the relative amounts of

Table I[14]

Personal consumption expenditure	*energy intensity (BTUs per dollar)*	*labor intensity (labor units per dollar)*
electricity	502,473	.044
gasoline and oil	480,672	.073
kitchen and household appliances	58,724	.073
new and used cars	55,603	.078
food purchases	41,000	.085
furniture	36,664	.092
women's and children's clothing	33,065	.100
religious & welfare activity	27,791	.086
telephone and telegraph	19,043	.055

Some expenditures — such as for clothing, food, and welfare activity — require relatively little energy while utilizing large amounts of labor. Energy-producing industries consume large amounts of energy, while creating very few jobs.

energy and the units of labor reflected in consumer purchases for 1971. Table I shows that for each consumer dollar spent on electricity, 502,473 British Thermal Units (BTUs) of coal, gas or oil were required. But only 44/1000 of a job unit was necessary. Gas and oil purchases reflect *similarly large energy requirements* per dollars spent by the consumer, and *relatively small* units of labor.

These figures contrast sharply with those relating to expenditures for other categories listed — such as women's and children's clothing, or furniture. These purchases needed far less energy than did production of electricity, gas or oil, and utilized far more units of labor per dollar spent.

In all, the major energy-producing and energy-using

industries consume a third of the nation's energy. Yet they directly provide only about 10% of the nation's jobs.[15] Energy companies claim that *indirect* employment created by energy is substantial. But as Professor Daly points out, *any* investment — even in welfare and unemployment compensation — leads to indirect job creation. And as noted above, energy, once available, *generally ends up replacing jobs.*

Where have the new jobs come from? Since World War II, new jobs have been created overwhelmingly in the merchandising and service sectors of the economy. Between 1947 and 1970, employment in these areas increased 95%.[16] These jobs have required relatively low amounts of energy, capital and resources. They also have caused less pollution and environmental disruption than industrial jobs.

The drive continues to substitute more and more energy for labor. The chief economist of the U.S. Department of Commerce recently testified before Congress that "productivity" will grow faster during 1976-1986 than in the past decade.[17] Indeed, "productivity" is becoming a factor in the service and merchandising sectors — the very areas which have provided new jobs in the past 30 years. Machinery is replacing supermarket clerks, bank tellers, teachers and telephone operators, in order to increase the output of remaining employees and avoid dependence upon human labor. The Bell System, for example, is planning to replace 33,000 operators by installing computerized switching and billing systems, at a net annual savings [to them] of $390 million.[18]

chapter 2 The Myth about Energy Expansion

Vast amounts of energy are wasted every day.
Energy conservation is the least expensive way to expand our energy "supply" — while creating new jobs.

The public has been led to believe that energy expansion is the springboard to economic growth, the "good life," and jobs. It's easy to discover who has promoted this myth.

Industry has been able to increase profits by replacing workers with energy — energy purchased at very low rates from an ever-expanding energy industry which has been accumulating ever-increasing profits. The small consumer has been picking up the tab: industries traditionally have paid *less* than individual consumers for each unit of energy used. In addition, by bearing most of the environmental and disease costs associated with energy, and by paying for substantial government assistance to energy companies, the public has actually been subsidizing industrial use of cheap energy to replace human labor.

It is thus clearly in the interests of both energy producers and major energy users to continue keeping the public in the dark about energy supplies, production, expansion and use. From their standpoint, it makes sense to dig up the old scare tactic about job loss and economic depression. The same line has been used with much success in the past: against the presidential bid of Populist William Jennings Bryan in 1896, against child labor laws, unemployment insurance, workman's compensation, labor union organizing, the minimum wage, and pollution control legislation, to cite but a few examples.

Critics of energy policy have been slow to challenge the industry-promoted mythology about energy expansion leading to economic prosperity and jobs. But increasingly, reputable organizations and analysts have been examining energy and its relationships to the rest of the economy:

• A number of studies have shown that Americans use twice as much energy per person as do West Germans, Swedes and the Swiss. Yet the standards of living in West Germany, Sweden and Switzerland are comparable to or higher than this country's. And unemployment in those European nations is much lower than in the United States. As a recent Federal Energy Administration report noted:

> This large disparity in energy use between [West Germany and the United States] suggests that continued economic growth and improvement of the standard of living in the United States should be possible without a proportional increase in energy consumption.[19]

• From his studies, Professor Herman Daly has concluded:

> Undeniably, energy is a necessary input to the supplies of nearly all sectors. But it does not necessarily follow that increasing the production of energy will increase the numbers of jobs in those sectors.... Nor does it always follow that other sectors cannot expand employment without an increase in energy input... In general,

> increasing input of energy to the supply side is neither a necessary nor a sufficient condition for increasing employment.[20]

• A research group in Washington, DC, Resources for the Future, has found that data

> do not support the widely-held belief that energy and Gross National Product (GNP) have grown at essentially the same rates in advanced countries.... It seems to be questionable to assert a rigid linkage between [the two].[21]

In the latter part of the 19th Century, says this organization, energy use in the United States grew faster than did the GNP. That was because heavy industry was the main factor in national economic growth. But after World War I, when lighter manufacturing plus increasing services began to predominate, energy use increased at a slower rate than did GNP.

• The Institute of Ecology, at the University of California at Davis, has concluded that comparison studies of different countries also

> do not support the hypothesis that constantly-increasing energy consumption per capita is a prerequisite to increasing economic growth.

The Institute discovered that energy and employment ratios varied in different industrialized countries, "reflecting fundamental differences between structures of economies." In addition,

> ...beyond an optimal level of energy consumption per person, further increases in energy use leads to a variety of deleterious effects which ultimately lower growth rates.[22]

• The Ford Foundation's massive Energy Policy Project determined that a lower rate of increase in energy expansion, accompanied by more efficient use of this energy, would expand the total number of jobs, and improve public health and environmental quality as well. By the year 2000, according to the report, improved energy efficiencies could actually permit a zero energy growth rate. In fact, the nation would be using almost half as much energy as is currently being projected for that date, and would be experiencing an increase in total employment.[23]

• An investigation by the Oak Ridge Institute for Energy Analysis has revealed that energy production could grow at the very low rate of 1.5% through the year 2010 without adversely affecting the economy. The Institute found that there would still be a real GNP growth of 2.5%-3%.[24]. According to *Science* magazine, this study provides "evidence that low-energy growth forecasts are well on their way to attaining the status of conventional wisdom."[25] And a study commissioned by ERDA commented that the data developed by the Institute represent

> a significant break with the proportionality between GNP and energy use which has held approximately even for several decades.[26]

• A recent report by the President's Council on Environmental Quality ("The Good News About Energy," February, 1979) has noted a number of studies which illustrate that

> the United States can do well, indeed prosper, on much less energy than has been commonly supposed. The principal basis for this good news is the accumulating evidence that the means are available to wring far more consumer goods and services out of each unit of fuel that we use.

• Among the works cited by CEQ was one prepared in 1978 by the Demand and Conservation Panel of the National Academy of Sciences' Committee on Nuclear and Alternative Energy Systems (CONAES). This panel of experts found that a "major slowdown in demand growth can be achieved simultaneously with significant economic growth." According to the President's Council, the CONAES data indicate that

> energy demand in the year 2010 could be roughly the same as today's level while providing a higher level of amenities, even with a total population increase of 35%.

• A report of the International Institute for Environment and Development (IIED) of London contends that a strong energy conservation policy would enable England to triple its Gross Domestic Product (GDP) with little or no growth in total energy demand. An analysis in the Environmental Data Services Report of 19 February, 1979, noted that the IIED study "rejects the notion that there is any hard and fast link between growth in GDP and growth in total energy demand," while it

> aims to dispel the myth that a lower energy future would necessarily mean fewer goods and services, fewer jobs and belt-tightening deprivation.

• John Myer, in the *Conference Board Record,* a journal which reports to management on business affairs, wrote that industry can use energy more and more efficiently. To Myers, the connection between energy growth and economic growth

> seems to me more elastic than commonly assumed. . . . To the extent that we can cut down on energy use more cheaply than we can expand domestic supplies, the goal of lessened vulnerability to supply interruptions or sudden drastic price hikes can be more economically achieved. This course would also carry the added benefits of less ecological damage and expense of environmental protection.[27]

• Testifying before the Joint Economic Committee of Congress, Lee Schipper and Thomas V. Long stated:

> There does not seem to be an immutable direct proportionality between economic growth and energy use.

Schipper and Long also pointed out that increased capital investment has been substituted for both energy and labor, hence

> energy use has been the tail of the dog, wagged by the interaction of the larger factor costs.[28]

• Even Paul McCracken, a former chairman of the President's Council of Economic Advisors, has confided to *Wall Street Journal* readers his doubts about what the public has been led to believe regarding energy and economic growth:

> the striking thing about the relationship between the rate of growth in output and the rate of increasing energy consumed is not that it obviously exists, but that it is so variable.[29]

Indeed, recent evidence, according to the New York *Times* (4 March, 1979) is that Americans consumed only about 1.9 percent more energy overall than they did in 1977, but the economy expanded 3.9 percent, or more than twice as fast. John Lichtblau, executive director of the Petroleum Industry Research Foundation and, according to the *Times*, a leading oil economist, is quoted as having stated:

> People were saying this kind of thing would lead to a stagnation of the economy, but it hasn't. It all makes sense now. It didn't [a few years ago].

"Americans," concludes the *Times'* analyst Anthony Parisi, "have learned to use energy more efficiently. In doing so they have loosened the bond that once seemed to tie energy and the economy so tightly."

Further examination of the myths surrounding energy requires discussions of energy waste, and of capital investment and job creation.

Energy Inefficiency and Waste

Individuals waste energy, and there is much each person can do to conserve. But the energy industry has tried to mislead the public into believing that the bulk of energy waste is caused by the average citizen who is too lazy to walk, or too pampered to lower the thermostat.

In fact, energy in this country is wasted primarily because public and private policies have long promoted energy extravagance in the mistaken belief that this is the best way to achieve a prosperous economy. The "lazy citizen" theory ignores the reality that accurate information on energy matters and alternative methods of energy supply have not been available to the consumer. It also does not take into account that wasteful energy consumption has been encouraged to increase the profits of energy-producing companies.

Jimmy Carter has stated:

> Our energy waste in transportation is 85%; in generating electricity, 65%. Overall, 50% of our energy is wasted.[30]

Denis Hayes, former director of the Illinois State Energy Office, and now a senior researcher at the Worldwatch Institute, has studied this waste in detail. Hayes found that in 1975, Americans wasted more fuel than was used by two-thirds of the world's population.

Energy can be wasted at three points: during generation; during transmission; and at end use. The process of transforming coal, gas or oil to electricity wastes about two-thirds of the energy in the fuel. Conversion of uranium to electricity is even more wasteful. Using coal to make synthetic fuels wastes about one-third of the coal's energy content.

Energy is wasted when it is sent out over long transmission lines, especially over power lines from nuclear power plants which for danger reasons the government has decided must be located far from large concentrations of people. Energy is wasted when appliances and machines are not built to perform at the best possible efficiency levels. Energy is wasted when heavy automobiles with "high performance" engines guzzle fuel to propel the vehicles a few miles per gallon. And energy is wasted when it simply is allowed to leak — through ceilings, walls, pipes, hot water heaters and industrial processes, due to inadequate insulation and design.

Energy is also wasted when it is not matched in "quality" to the intended task. This occurs, for example, when electricity is used for heating buildings and water, and for cooking. Electricity is a very "high quality" and expensive form of energy, most suited to and most economic for special tasks such as rail transportation and some exacting industrial processes. When water is heated by electricity, for example, it is after water at a power plant has already been boiled by the fuel. This fuel, whether fossil, nuclear or solar, has had to boil water to make steam to turn turbines to generate electricity — and in each step there is considerable loss of energy. It is quite apparent that there is a wasteful "mismatch" when nuclear fuel is used to achieve a temperature of several thousand degrees in an expensive and complex reactor in order to boil water to create steam — something which occurs at 212 degrees F. And it makes even less sense to go through all this trouble to heat or cool buildings to 70 degrees when other sources of less troublesome and less expensive heating and cooling are available which utilize air whose temperature is closer to the levels desired. For instance, modern high-rise office buildings could be well served by heat pumps which move cold air to the warm side of the building exposed to the sun's rays, and return warm air back to the cold side. These heat pumps would replace the waste created by heaters and air conditioning units operating simultaneously in the same building.

Energy is wasted at industrial sites when steam, which could be reused to generate on-site electricity or put to a variety of successively lower-grade uses ("cascading") is simply released to the atmosphere. Energy is wasted when inaccurate long-range projections lock utilities into building huge, central-station electrical generating plants which are not needed when they are completed seven to ten years later.

According to Amory Lovins, if we increased electrical production by the year 2000 to the 150 quadrillion BTUs which many energy and government officials yearn for, it would result in wasting as much energy as the U.S. consumed in all of 1971.[31]

And as energy is wasted and thrown away, so are jobs. Bruce Hannon has computed that for each quadrillion BTUs of fuel changed into electricity, 75,000 jobs are lost over the entire economy.[32]

Finally, all the money invested in wasteful and unnecessary energy production at the expense of other national needs must be considered wasted as well. This is especially serious because money will be in short supply: Lovins doubts whether the nation will even be able to afford to build — or consumers able to buy — all the products which are supposed to be using this wasteful investment in energy.

Capital Investment

The major energy-producing and energy-using industries consume the lion's share of national capital investment.

A recent ERDA report concluded:

> It is probable that the nation's single greatest investment in energy in the future will be in the area of electricity generation, transmission and distribution.[33]

Amory Lovins has calculated that investments in large-scale, centralized energy systems as proposed by people in DOE and assorted energy monopolies would consume three-fourths of all private investment capital in the United States.[34] Estimates of the total amount of money needed through 1985 range from $900 billion to one trillion dollars.[35] The Edison Electric Institute, the public relations arm of the nation's privately-owned utilities, has estimated that these utilities would require $570 billion for construction purposes alone to provide electric power they believe will be necessary over the next fifteen years.[36] The President of Exxon Nuclear Corporation stated a few years ago that at least *half a trillion dollars* would be needed to build the 500 nuclear power plants he would like to see completed by the year 2000. Another hundred billion dollars, he added, would be required to construct the necessary nuclear fuel cycle facilities.[37] The development of a commercially-usable nuclear breeder reactor, by the latest estimate, will cost $10 billion plus. Storage of radioactive wastes will consume presently-uncalculable billions of dollars over hundreds and

Table II[38]

Capital investment per employee	*Industry*
$108,000	petroleum
105,000	public utilities
41,000	chemicals
31,000	primary metals
24,000	stone, clay, glass
19,500	all manufacturing (average)
18,000	food & kindred products
11,000	textile mill production
11,000	wholesale and retail trade
9,500	services
5,000	apparel & other fabricated industries

Capital investment per job: the same investment would create 10 to 20 times as many jobs in the textile industry, service sector or retail trade as it would in public utilities or petroleum.

hundreds of years (uncalculable because the methods for long-term storage have not yet been devised). The cost of each coal-conversion facility is currently thought to be around $1 billion. Rocky Mountain shale conversion plants will run at least $1 billion each. And the price tag for nuclear fusion electric systems, assuming they can be made commercially available, will be in the billions of dollars.

These large capital investments result in small numbers of very expensive jobs. A 1977 report by *The Conference Board* found that investment to create each job in energy-producing and major energy-using industries was much higher than investment to create a job in areas which use smaller amounts of energy. See Table II.

It is clear from these figures that it takes about 21 times the amount of investment to create a job in the petroleum industry as it does to create one job in the apparel and textile industry. Jobs created by investment in public utilities are the second-most expensive: about $105,500 each. The investment needed to create a manufacturing job on the average is about $19,500, which is about *one-half* of the amount required for a job in the chemical industry, and about *one-third* of what is required per job by the primary metal industry. Investment in "all manufacturing" is about *double* that needed to create a job in wholesale and retail trade, or in the service sectors of the economy.

But despite the huge capital investments required, and the small numbers of jobs created; despite the inefficiencies and the waste of non-renewable fuels; and despite threats to human health and local autonomy, the energy industry continues to promote its large-scale, complex energy systems. (In 1976, for example, over one-half of all new homes constructed had all-electric heating, cooling and cooking systems.) In attempting to fulfill their own propaganda that prosperity and jobs depend on more and more energy — especially electrical energy — these companies will seek to divert more money and more resources their way to create more waste and more pollution.

The trend will continue toward fewer manufacturing jobs, greater gaps in wage scales through the entire society, increased environmental destruction and citizen protest. Less money will be available for investments in other areas which serve people's needs, provide more jobs per dollar, consume less energy, and create less environmental and public health havoc. Consumers will be paying more for unneeded, inefficient and unreliable energy systems which will be unable to provide a return on the money invested in them. Consumers will also be paying higher taxes to protect their health and to bail out failed energy systems. (This is happening now: Nuclear Fuel Services' fuel reprocessing plant and a Con Ed nuclear reactor in New York State have already been dumped into the laps of the people of New York.) Higher interest rates will also result because of the money shortage caused by the diversion of so much capital to the energy arena. (Some projections show this shortage may total $915 billion through 1985.[39])

Surely, there is a better path to prosperity and jobs.

chapter

3 Jobs from Conservation

A modest investment in energy conservation would create a wide range of jobs. Energy conservation just in buildings would create at least half a million direct jobs through 1990.

Energy production is not a goal in and of itself. Energy should be utilized to serve people, to provide the freedom for all people to have richer, easier, healthier lives. That a nation uses vast amounts of energy does not mean that the energy is actually being put to wise, effective use by its people.

The best approach to energy sufficiency, economic prosperity and jobs is that which combines increasing energy efficiencies with a variety of diverse and safe energy-supplying technologies. Each energy-producing technology should be used to do what it does best, and should be matched in scale and energy quality to the way in which its energy will be used. And the more the fuels for these new energy systems are renewable, the better.

This approach is not "anti-technology," as sometimes is alleged by the large energy interests. In fact, technological innovation will be a key to achieving success with this approach... but the technologies involved need to be ones which can be controlled by the American people, not ones so elaborate and complex that people have to be kept far away from them and from decisions concerning them.

And this is not a "no growth" approach, or one which advocates a return to drudge labor. To the energy industry, "growth" has always meant growth in energy production in order to satisfy its own needs, no matter the consequences for the rest of society. But to others, "growth" means a national policy of full employment, improved standards of living, improved job safety and public health, expanded opportunities for leisure activities and the development of rewarding relationships with other people.

How can increasing energy efficiencies (that is, conservation of energy) make a significant contribution to such goals? Here is a brief list of just some of the benefits which result from energy efficiencies:

• To the University of California's John Holdren, the essence of conservation is "extracting more well-being from each gallon of fuel and each kilowatt of electricity."[40]
• A unit of energy saved is cheaper — in dollar, energy, public illness and environmental costs — than a unit of energy produced. The Bonneville Power Administration has concluded that saving energy by conservation costs *one-sixth* of what it costs to deliver an equivalent amount of energy from new thermal power plants.[41]
• "Saving energy," said Frank Zarb, formerly director of the Federal Energy Administration, "is synonymous with saving dollars and can, in fact, be one of the least expensive energy supplies this nation has."[42]

Chart I[44]

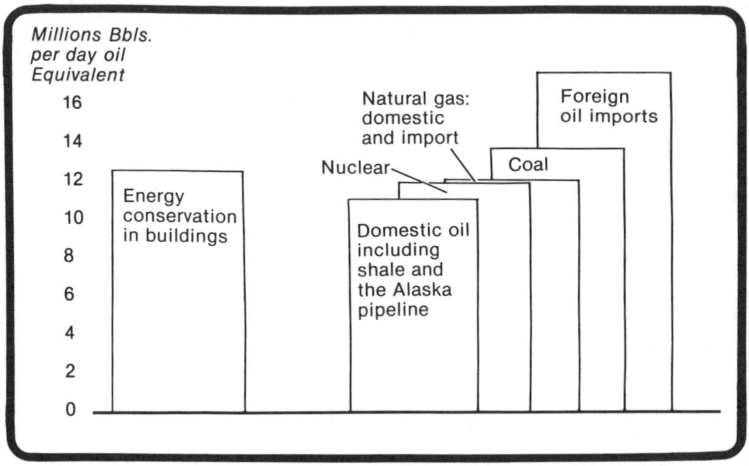

Energy conservation *in buildings alone* could, by 1990, create a larger energy supply than would domestic oil, nuclear energy, or domestic and and imported gas. (Based on U.S. government forecasts of potential production and savings.)

• Denis Hayes has calculated that for the next quarter century, the United States could meet all its new energy needs "simply by improving the efficiency of existing use."[43]

• The American Institute of Architects (AIA) has calculated that by 1990, 12.5 million barrels of petroleum per day (equal to one-third the current national energy use) could be saved just by employing energy-efficient systems in old and new buildings. Chart I, prepared by the AIA, shows that conservation alone in buildings offers an energy supply which by 1990 would be larger than the contribution of any one of the following: domestic oil, conversion of shale to oil, Alaskan North Slope oil, domestic and imported natural gas, and an overly-optimistic prediction of nuclear energy output.

In order to supply the energy equivalent of 12.5 million barrels of oil per day in 1990 by traditional centralized energy systems, a capital investment of $415 billion would be required. The customer would pay between $892 billion and $1.4 trillion for this unneeded energy. But investing money in energy efficient buildings, says the AIA, would produce both energy and dollar savings.

• The Bonneville Power Administration has summarized a number of energy-savings estimates calculated by its consultants and a variety of other analysts. See Table III.

• General Electric's Evandale, Ohio jet engine plant has cut its energy use by 45.1% over the last five years, "with nothing more sophisticated than a chicken coop timer." G.E.'s program has saved the company $8.6 million, at an investment of $400,000.[46]

• 1976 cars already average 27% better mileage than 1974 cars. 1977 cars will average 6% better mileage than the 1976 models.[47] Congressionally-imposed mileage standards are forcing the auto industry to improve energy efficiency of their cars every year, despite the opposition of the giant car makers.

• An energy study commissioned by the City of Seattle influenced that city not to buy into Oregon's nuclear power plants. The study convinced city officials and Seattle citizens that with appropriate energy efficiency measures, *no new electricity generating capacity would be required for Seattle through 1990.* And, the cost per kilowatt hour of electricity would be 1.3¢ *cheaper* without building new energy facilities.[48]

• A Dow-Midland study indicated that by using waste industrial steam to generate electricity — as is done in Sweden and

Table III[45]

BPA Study (1995 savings)	*Other studies*	
Residential Sector		
14-48%	24.8%	Energy 1990 (City of Seattle, 1976)
	38-49%	U. of Calif. (Goldstein, 1975)
	30-40%	Oak Ridge National Lab. (1973)
	11.3-42.7%	Arthur D. Little, Inc. (1976)
Commercial Sector		
22-44%	30%	Energy 1990 (City of Seattle, 1976)
	30-80%	American Institute of Architects (Stein, 1974)
	30-50%	U.S. General Services Adm. (1973)
	59.6%	Arthur D. Little, Inc. (1976)
Industrial Sector		
5-10%	31.7%	Energy 1990 (City of Seattle, 1976)
	24%	West German Industries (Stanford Research Institute, 1975)
	15-30%	National Bureau of Standards (1974)
	30%	Federal Energy Adm. (1974)

Numerous studies have demonstrated the vast potential for cutting energy waste, especially in housing and commercial sectors of the economy.

West Germany — energy savings equivalent to 680,000 barrels a day of oil could be saved by 1980. By 1985, this "co-generation," as it is called, could replace the equivalent of 50 large nuclear reactors.[49]

• The California Energy Commission has determined that the potential for co-generation in that state alone could be as much as 140 billion kilowatt hours per year, the equivalent of the total amount of electricity consumed in California in 1975.[50]

• Energy conservation retrofits have led to a 60% decrease in steam use, and a 30% decrease in electrical consumption, at the Hermann Building of the Massachusetts Institute of Technology.[51]

• The Director of the Massachusetts Energy Policy Office has concluded:

> Energy conservation is not synonymous with deprivation. In fact, just the opposite is true. Saving energy means saving cash. For a household, this means more cash for other activities; for a business, this savings means higher profit. For the state in general, this means more jobs, and more income for our citizens.[52]

• States have responded to the Energy Policy and Conservation Act (1975) which provides government money to a state which can cut back at least 5% of its energy use projections, beginning in 1980. Colorado plans to save 52.8 trillion BTUs in 1980. Governor Lamm, in his state's energy conservation plan, sums. up the rationale:

> Conservation is the only way in the next few years we can bring some manner of economic relief to the millions of our citizens who cannot afford the vastly increased cost of energy. Conservation is the least expensive and most cost-effective way to expand our energy "supply"... A conservation program will reduce the impact [of energy production] on the environment...
>
> An essential part of any serious program to promote energy efficiency must be a systematic and credible effort to convince the public of the enormous benefits that are possible... The economic and energy savings can be calculated, and they are attractive.[53]

• Again, FEA's Zarb:

> Contrary to myth, conservation is vital to our efforts to sustain our high standard of living and rekindle economic growth. Moreover, several recent analyses have shown that reducing the inefficient use of energy would not result in an employment penalty.[54]

• And even Americans for Energy Independence, a group oriented to the energy industry and seeking the expansion of large-scale, centralized energy systems, has noted that:

> . . . energy conservation and economic well-being are compatible. . . Even the most superficial and immediate measures to cut waste would revive our environment.[55]

• Conservation of energy will supply goods and services at lower total cost than could be supplied by new energy development. The money which might have been spent on energy production would be available to pay for labor, designs, and materials for energy efficiency techniques. There will also be money left over for other investments. The result will be more goods, more services, more employment — both direct and indirect, as the Bonneville Power Administration has pointed out. And, as Denis Hayes has observed,

> To the extent that a consumer conserves fuel and spends the money on anything else, he will provide more employment as well as use less energy.[56]

• Energy conservation offers considerable manufacturing flexibility. Most energy conservation technologies will be implemented on small scales, and on regional bases. Industry will therefore be able to develop improved technologies through competitive trials, then cut production costs by means of mass production. A "balanced" and independent regional economy based on improved energy efficiencies could provide more jobs, with more job security, than an economy dependent upon energy sent in from far away. The reason for this is that the manufacturing base would be varied, would require less energy, and be less sensitive to changes in economic, political and meteorological conditions in other parts of the country.

In addition, raising investment capital would be less of a problem because energy efficiency programs are cheaper than energy-production projects. Their short planning and completion time would mean that cash flow would be speeded up: money for energy efficiency investments could be used several times rather than be tied up in a long-term energy production project.

And with regard to jobs created by energy conservation, there are significant positive benefits as well.

Energy Efficiency and Jobs

Energy conservation creates jobs.

• A Bonneville Power Administration Study has found that

> High impact conservation programs create more jobs than would be created by building new power plants to generate an equivalent amount of energy.[57]

A study published by the Congressional Joint Economic Committee, "Energy and Economic Growth," by Marc Ross·and Robert Williams, (1977) determined that an expanded effort to increase energy productivity would "create many new job opportunities and encourage productivity gains" throughout the economy, according to the CEQ's "The Good News About Energy."

Amory Lovins has testified in the Senate that conservation programs which shift investments from energy wasting to social programs create from tens of thousands to nearly a million net jobs per quadrillion BTUs of energy saved.[58]

• A preliminary analysis for the FEA[59] provides specific breakdowns of some energy conservation techniques, costs and resulting employment. This report examined the prospects of limited energy efficiency increases in 34,372 private homes. The technical work called for was simply the installation of ceiling insulation and automatic thermostats, and the retrofit or replacement of furnaces.

The analysis concluded:

> By 1985, natural gas supply would be increased because of the saving of 1,212 billion cubic feet. This is the equivalent of the gas to be obtained from the major discovery at the Alaskan North Slope. It is also about the equivalent of the output of 39 one-thousand megawatt electric thermal power plants. Consumers in these 34,372 homes would save \$1.7-\$2.3 billion in heating costs.
>
> The work would cost \$7-\$10 billion, compared with \$17-\$20 billion for 39 large fossil fuel power plants; 487,000 jobs over 7 years would be created: 122,000 in manufacturing, 366,000 in local installation.

The report also stressed that employment associated with energy conservation techniques is local, low- to moderately-skilled, and concentrated in or near urbanized areas which are experiencing the most acute unemployment problems. In contrast, centralized, expensive energy production complexes usually have to bring in highly-skilled labor from outside the construction area. (These transients create a large amount of disruption: temporary housing and many services must be supplied to meet the problems temporary workers create. In many of the energy "boom towns" of the Western United States, crime, alcoholism, and family break-ups are well above average. Serving the needs of transient labor ends up being a drain on the local economies the transients were supposed to be stimulating.)

• The State of Colorado has estimated that 17 million private homes in the nation need ceiling insulation; 20 million need clock thermostats; 20 million need caulking and weatherizing; 10 million need storm windows. In addition, millions of multiple-unit dwellings, apartment houses, commercial and industrial buildings are inadequately insulated and weatherized. If 487,000 jobs are generated by only three simple conservation procedures in a small fraction of the buildings which need work, the potential obviously exists for millions of jobs. Employment for energy specialists, construction of more energy efficient equipment, jobs for architects and engineers will be increased, along with jobs in the fields of insulation, heat pumps, electronic controls and systems analyses, communications and transportation.

• The Senate Commerce Committee Staff has estimated that $1.6 billion in interest subsidies and loan guarantees for conservation retrofits would generate 400,000 jobs.[60]

• The American Institute of Architects' conservation program for new and existing buildings would create half a million to more than a million direct jobs per year through 1990.[61]

• The conservation retrofits on MIT's relatively small — 94,000 square feet — Hermann Building, provided 800 person hours for members of the Sheet Metal Workers' International Association.[62]

• A proposal sent to President Carter by the Senate Commerce Subcommittee Staff describes how a $1.65 billion investment in energy conservation, using public service workers, would create a total of 100,470 jobs: 9,600 at $15,000 per year; 74,600 at $10,000; and 16,279 jobs in manufacturing. The staff calculated that each $40,000 investment in materials would create one job in the materials manufacturing industry. The program would focus on the retrofitting of schools, colleges, hospitals, federal buildings, the weatherizing and installation of solar water heaters in HUD-owned single family houses, and the construction of bicycle paths.[63]

• A study by Skip Laitner of *Critical Mass Journal* computed that manufacturing high efficiency appliances requires the employment of more people than does the manufacturing of inefficient appliances.[64] When buying more efficient items, a consumer is investing in extra labor and materials, and in a more carefully-constructed, longer-lasting product. The purchaser would save money on reduced electric bills, which would cover the higher initial cost of the appliance. Thereafter, ongoing savings in energy can be spent on consumer goods, which as described above, create more jobs than spending on energy generation.

• According to *Americans for Energy Independence,*

> experts have suggested that reducing oil imports by 2 million barrels a day would generate 500,000 to 800,000 new jobs in the U.S.[65]

• And we could save 144 trillion BTUs a year by switching to refillable-container systems. That would result in a net job increase of 117,000. Total labor income would increase $936 million, according to a recent FEA study.[66]

In sum, increasing energy efficiencies will supply the nation with a substantial source of energy. The savings of money and the reduction of energy waste will generate a broad range of economic and environmental benefits throughout the country, and create large numbers of safe, socially-useful jobs.

For *new* energy sources, the nation needs to utilize the unlimited energy which is available from the sun.

chapter 4 Jobs from the Sun

The technology for a solar energy economy is already here. The transition to solar systems would create vast numbers of jobs in construction, engineering, and other fields.

The solar energy reaching the United States in twelve hours is equal to the nation's yearly energy consumption.

Various technologies exist which can utilize this abundant and safe source of energy: solar plate collectors for heating and cooling of buildings, and for heating water; photovoltaic cells for converting sunlight directly into electricity; wind energy systems for producing electricity, pumping fluids or compressing air; biomass conversion for creating fuels from organic matter.

For all solar energy systems, the fuel is renewable. For all except biomass conversion, the fuel is free. If energy quality is matched to its ultimate use, each of these systems except photovoltaic conversion has lower capital requirements per unit of energy delivered to the user than does nuclear fission and production of synthetic fuels from coal. All these solar technologies readily lend themselves to small, decentralized installations.

According to John Quarles, former Deputy Director of the Environmental Protection Agency, some solar energy techniques can be commercialized sooner than can the nuclear breeder reactor, shale or coal conversion systems, or nuclear fusion reactors.[67]

Senator Charles Percy (R-Ill.) recently stated at a solar energy conference he helped organize: "Solar energy is not an exotic dream. Rather, it is workable today."[68]

Some solar energy devices are already on the market. The Grumman Corporation "Sunstream" division has advertised: "An answer to OPEC comes up every morning!" With Sunstream's domestic hot water heater,

> you can now capture the sun's energy to heat hot water for your household needs... efficiently, quietly and cleanly.[69]

An ERDA report, *A Preliminary Social and Environmental Assessment of the ERDA Solar Energy Program,* has pointed out that solar energy technologies are readily adaptable *today* for broad scale commercial use.[70]

According to the old Federal Energy Administration,

> In general, the application of solar energy techniques does not require establishment of scientific feasibilities.[71]

George Szego, president of Intertechnology Corporation, has stated that savings of five billion barrels of oil could be achieved within 10-15 years if a crash program of solar energy commercialization were begun now.[72]

Edward J. Carlough, General President of the Sheet Metal Workers' International Association, recently noted that almost

25% of all energy used in this country is consumed in heating, ventilating and air-conditioning, and related systems. He went on to say:

> At a relatively low cost, we could make a huge dent in the nation's energy consumption by modifying those systems to use solar energy where it is feasible. The national value of a concerted effort in solar energy and energy conservation would be measured in millions of barrels of oil — resulting in more favorable balances of payments, less inflation, more jobs and energy independence.[73]

The Sheet Metal and Air Conditioning National Contractors' Association has testified before Congress:

> We have the skills, shop facilities, manufacturing capability and expertise to embrace this industry now. New tools and processes are not required.[74]

A report prepared in *1952* by the Paley Commission for President Truman concluded that solar energy could play a greater role in energy production than could nuclear fission. The Commission projected that if an aggressive effort were made, 13 million homes and commercial buildings could be heated by solar energy by 1975.[75] And the Atomic Energy Commission concluded in 1974 that by the year 2000, solar energy could provide 30% of the country's energy needs.[76]

Although efforts to enable these predictions to come true were not made, there is nonetheless a significant amount of solar activity in this country today. The Solaron Corporation of Denver had a half-million dollar order backlog in 1977, and was installing hot water systems in 60 buildings in 12 states.[77] The Piper Hydro Company installed 1000 dwelling hot water units in 1976, and "hasn't even scratched the surface" of demand.[78] The Federal Energy Administration has announced that commercial production of solar collectors increased 400% in 1975 over 1974. A total of 142 firms are now manufacturing collectors, and the industry seems to be in a trend of quadrupling production annually.[79] Solar energy businessman Jerry Plunket has estimated that 100,000 shops, manufacturers and firms are capable *now* of producing reliable solar energy equipment.[80]

Solar energy studies — even those which suggest that a significant contribution by solar technologies to the nation's energy needs is a long way off — constantly stress how much energy could be obtained from the sun, and how safely. FEA's former director Zarb says: "Solar energy appears to be the safest and most environmentally sound of all energy. . . . It is inexhaustible, and it can't be embargoed."[81]

The President's Council on Environmental Quality, in a 1978 report titled "Solar Energy: Progress and Promise," announced that with a program of accelerated development, solar technologies could be supplying 25% of the nation's energy by the year 2000, and 50% by 2025.

"Distributed Energy Systems in California's Future," a 1978 analysis done for the Department of Energy by the Lawrence Livermore Laboratories and the University of California at Berkeley, has found that by 2050, California could be energy independent, with a vigorous economy based primarily on renewable energy sources.

A study by the Oregon Energy Council found:

> A transition to a solar energy economy is desirable and realizable. It involves neither privation nor social deprivation. Life-style changes would be minimal. The rewards would be enormous. Our children would have a totally indigenous, permanent, safe energy system which could be relied on by countless generations for the future.[82]

The ERDA *Social and Environmental Assessment* of solar energy concluded:

> Solar energy is the one source of energy for which there are no fundamental obstacles, no insurmountable barriers, no serious environmental problems, no organized public interest opposition.[83]

The Solar Task Force of the Project Independence Report made clear that:

> A major shift to the utilization of solar energy by the year 2000 and beyond will reduce the effects of rapidly-growing energy consumption on our global climate through a reduction in heat waste rejection at the generating site. Solar energy is the only energy source which presently holds this promise.[84]

And even according to the Exxon Corporation, a multinational company heavily invested in oil, coal and uranium, and which is fond of predicting that solar energy technologies will not be particularly useful until the next century, the sun is:

> a source of energy that doesn't in any way pollute the air or water, can't be used up, and is found right in our own backyard.[85]

In brief, the greater the experience which is reviewed, the more analysis which is undertaken, the more optimistic become the judgments that the various solar technologies can indeed make a *major* contribution towards fueling our economy — and much more quickly than was thought to be possible just a few years ago.

Here is a brief look at some of the ways solar energy can be utilized to meet the nation's needs:

Solar Heating and Cooling — Solar heating and cooling uses collectors on site to absorb the sun's energy. This method of heating buildings and water is not new. In fact, it has been used in this and other countries for many years.

According to former Congressman Paul Cronin, solar heating and cooling systems

> can have an immediate effect on our overall energy budget... Most of the U.S. patents on solar [heating and cooling] were issued in the 1920's. Florida received the majority of its hot water from solar energy until World War II... If the government decided to provide a solar hot water heater to everyone in the U.S., which could be done in 3 years, it could save over 1.2 million barrels of oil per day, and create... jobs that would pay taxable incomes.[86]

Again, Sheet Metal Workers' Association President Carlough:

> Solar heating/cooling systems using air are no longer a novelty. This is a technology capable of producing energy and jobs.[87]

The Technical Development Corporation (TDC) of Providence, Rhode Island, designed the solar heating system for the swimming pool at the United Auto Workers' (UAW) Family Education Center in Northern Michigan. The installation costs for the UAW system will be paid back in 6½ years. Thereafter, $3000 per year will be saved in decreased bills for conventional fuels. TDC studies have confirmed that the use of solar heating for swimming pools, laundries and industries which use large amounts of water, will generate similar rapid returns.[88]

The Energy Research and Development Administration concluded two years ago that solar residential hot water and space heating were economically competitive with electric systems in 12 geographically-dispersed American cities, including Boston, New York, Madison and Washington, D.C. These

Chart II[90]

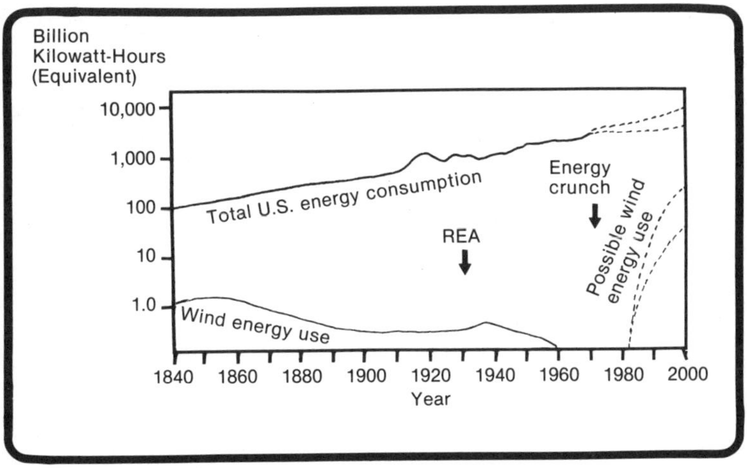

Use of wind power has dropped drastically in the last century.

kinds of solar units, stated ERDA, could become economically competitive with "most fuel alternatives" by 1980.[89]

Wind Energy: Research and development in wind power have been virtually non-existent in this country for about 30 years. But since the mid-19th century, more than 6 million small windmills have been used to pump water, compress air and generate electricity. Over 150,000 are still in operation. Significant use of wind power was phased out in the 1930's when the Rural Electrification Administration (REA) introduced centralized electric power, as shown in Chart II.

According to a 1972 National Science Foundation/National Aeronautics and Space Administration report, wind energy systems could be installed in a very short period of time — less than two years. Companies such as Boeing, Kaman, General

Electric and Lockheed, along with a growing number of smaller firms, are willing to provide wind energy systems this rapidly.

Wind power expert Loren Johnson has testified:

> Wind energy systems not only give a utility [company] more flexibility than do conventional thermal power plants due to shorter lead times, they also allow new technological innovations to be easily and rapidly incorporated as they occur... The failure of one wind energy system will not force the shutdown of another... The failure of one wind energy system in a large array is... not a major megawatt catastrophe.[91]

The Project Independence "accelerated development" scenario, which many experts consider to be quite conservative, estimates that 23% of the nation's electricity could be generated by wind power by the year 1990.[92] And a National Science Foundation-Mitre Corporation study found that electricity produced by wind power would be competitive with conventional diesel fuel electrical generating plants which use oil priced at $10-$11 per barrel, a price already consigned to days gone by.

The Technology Development Corporation has calculated that wind generation of electricity is currently economically competitive with other systems on the seacoast of the Northeastern states. As complementary energy systems, they are "immediately viable." The wind machine this company built for the United Auto Workers will shortly be producing power in cooperation with a Northern Michigan utility.[93]

In northern New York state, the Energy Research and Development Authority has installed a demonstration wind generator on a working dairy farm. The Authority's chairman said: "Cost-effective use of new [wind] energy technologies is available today."[94]

Biomass Conversion: Biomass conversion means changing organic matter into useful fuels, such as gas or oil, or directly into heat.

The various biomass techniques include utilizing both urban and industrial wastes, agricultural and forest residues, along with farming on land or in the ocean.

According to Senator Peter Domenici (R-New Mexico), the advantages of biomass conversion are many: sources are forever renewable; the environmental impacts are minimal; marginally useful lands and waters can be used; a range of assorted bioconversion products, which differ in kind and quality, can be matched to physical and economic needs.[95]

Substantial amounts of energy can also be saved by recycling paper, metal, glass and plastic fractions of the municipal solid waste stream into the secondary materials industries. Manufacturing processes using these materials require less energy than when virgin materials are used. These recyclable materials comprise about half of our municipal solid waste. In addition, about 30% of solid waste (household organics plus yard trimmings) can be composted to produce a variety of end products which can reduce the need for oil-based fertilizers. If given appropriate attention and investment, the recycling sector of our economy can expand greatly within a very short period of time. Non-recyclable fractions of the solid waste stream can be burned to produce energy in appropriately-scaled energy recovery systems.

Harvest and bioconversion of ocean kelp into methane gas would generate 23 trillion cubic feet of gas per year. (This is enough to completely meet the current national demand.) Land plantations could generate 8-11 trillion cubic feet of gas per year. Some gas companies are presently experimenting with ocean- and land-cultivated kelp production.

China has built one-half million methane gas digesters since 1970. India is constructing 100,000 of them.

Bioconversion techniques are especially conducive to dispersed, flexible, low-capital enterprises which create many jobs. The head of the American Gas Association has said:

> From a social or political standpoint, it appears desirable that gas be generated on a local basis from such sources as urban refuse, industrial wastes, sewage sludge and miscellaneous wastes.[96]

Photovoltaic Conversion: Solar photovoltaic cells convert sunlight directly into electricity at the place where the electricity is to be used. They are already cost-competitive with other energy systems in remote areas. Their efficiency has been going up, and their price going down, despite limited federal and industrial interest. In the early 1960's, energy from photovoltaics cost about $200 per peak watt. By the early 1970's, the cost had dropped to $100. By 1973, it was $30; and in 1975, $17. Recently, the energy from photovoltaic cells has been available for $10 per peak watt. A 1977 study from the Jet Propulsion Laboratory and the California Institute of Technology has indicated that the price will come down to 50 cents per peak watt by 1984.

Since photovoltaic cells convert solar energy to electricity on site, there is no need for long-distance transmission lines, and therefore no energy loss along these lines. Solar cells, like wind systems, can be installed in small units, and added on to when needed. Photovoltaic cells don't require energy planners to predict years in advance what the specific energy demand of a region will be.

A 1975 National Science Foundation report found that by the year 2000, photovoltaic electricity could supply 9% of the nation's needs — a figure which many today consider understated.

The evidence is overwhelming that solar energy technologies are available today, and that they offer many technical, environmental and financial advantages over other energy systems.

They also provide significant employment benefits.

Solar Energy and Jobs

Douglas Fraser, president of the United Auto Workers, explained in an April, 1978, publication called "Worklife," published by the U.S. Department of Labor, why his union was supporting Sun Day organizations and celebrations around the country: "Today, when our nation needs both safe energy and good jobs, solar power can provide both."

Frank Zarb has said:

> A strong solar industry means more jobs for Americans, who will be needed to design, manufacture and install and maintain solar equipment.[97]

The Sheet Metal Workers' President Carlough:

> . . . even figured conservatively, energy-saving modification work and an expanded use of solar energy could put all unemployed sheet metal workers back to work.[98]

The Massachusetts Energy Policy Office recently concluded

that if by 1985 one-half of all buildings in that state were to use solar energy for hot water production, 32,000 jobs would be created. In addition, 600 million gallons of oil would be saved per year, and $480 million retained for spending within Massachusetts. The report stated:

> The potential for solar energy systems seems virtually unlimited. With widespread adoption of solar power, Massachusetts citizens could cut their collective fuel bills by $120 million annually by 1985. Furthermore, solar energy has vast potential for new job opportunities, especially in the plumbing, construction and research and development areas... It's safe to say that by 1995 more jobs could be available from solar power (directly and indirectly) than from offshore oil and new nuclear construction combined.[99]

Robert Mills, of the Sheet Metal and Air Conditioning Contractor's Association, has reported that retrofitting just 3 million private homes to 60% reliance apiece on solar energy for heating and cooling would lead to 12.2 million hours per year of work for 10 years. And putting solar heat in 2.3 million new homes, at 60% reliance, would create another 12.2 million hours per year of work for ten years... for sheet metal fabricators, sheet metal installers, asbestos workers, carpenters, plumbers and pipefitters. Increasing the number of installations to half of all private homes would create about 5 times as many jobs. In addition, thousands of jobs would be created in retrofitting and new installations of solar systems in commercial buildings, apartment houses and government buildings.[100]

A 1975 study for the Sheet Metal Workers' International Association found that heating and cooling could provide a $2 billion market annually by 1990. Installation labor would cost half a billion dollars, of which $240 to $300 million per year is expected to be paid to sheet metal workers (about 15,000 jobs at $20,000 per year).[101]

The Laborer, a journal of the Laborers' International Union (AFL-CIO), found that jobs for its members in the solar energy field "could well mount into the hundreds of thousands." The union has begun a course in San Diego to train laborers in the installation and maintenance of solar and wind systems. Union President Angelo Fosco has said:

> Experts estimate the annual market for installing solar systems and converting existing structures to solar systems has a potential of $77 billion alone. . . . not including maintenance. . . . That translates into a goodly number of jobs for construction workers in our jurisdiction.[102]

Former president of the International Association of Machinists and Aerospace Workers (IAM), Floyd Smith, told delegates at an IAM-United Auto Workers legislative conference in 1975:

> If, for example, the government launched a program tomorrow morning to equip each home in America with a rooftop solar water heater, scores of factories would be retooled and reopened. Thousands of jobs would be created for unemployed machinists and auto workers.[103]

The IAM Grand Lodge Convention in Florida passed the following resolution in 1974:

> An all-out federally-sponsored program to convert home water and space heating to solar would not only decrease America's dependence on foreign oil, but create millions of needed jobs for construction workers, machinists, metal workers and other industrial workers.[104]

The IAM newspaper, *The Machinist,* quoted one solar advocate as saying: "Using local solar energy makes much more sense than carting oil halfway around the world." And

Machinist associate editor Dean Ruth wrote:

> Millions of Americans are jobless. Other millions face the threat of unemployment. Solar energy, as a new industry, can provide a vast amount of work to help get America moving forward again.[105]

An economist with the International Woodworkers of America (AFL-CIO/CLC) has written:

> Employment opportunities exist with the development and construction of alternate energy sources such as solar, tidal, geothermal and wind. . . . sound conservation methods and the development of alternate [to nuclear] sources will not adversely affect job opportunities in the economy. The job impact would, in fact, be positive.[106]

Another article in the Labor Department's "Worklife"[107] was titled "Solar Energy: Potential Powerhouse for Jobs." It listed some of the skilled workers needed to build and maintain solar units: carpenters, cement masons, electricians, plumbers, sheet metal workers, air conditioning, heating and refrigeration technicians, glaziers, crane operators. New jobs, the article says, will have to be created and ground broken in solar engineering, architecture, law, real estate and appraisal, sales, zoning, assessment and consumer protection.

The FEA Project Independence Task Force found that 3 to 4 million person-years of direct jobs would be needed in solar energy development and operation by 2000.[108] This figure is probably an underestimate, since FEA's 1974 "accelerated" rate of solar development is thought by analysts today to be too conservative. Among other things, it is based on oil selling at $11 per barrel when it is now selling for as high as $20 — and going up; also, some of the solar technologies were considered for only certain parts of the country, i.e., the Southwest, which many believe an unnecessary limitation. Dr. Jerold Noel, for example, a physicist at Mobil-Tyco solar labs, has stated:

> The roof of an average house in Philadelphia could produce enough energy to supply the needs of a home, with enough energy left over, say, to charge an electric car.[109]

The jobs, according to the Federal Energy Administration, would be for the skilled tradespeople listed in *Worklife,* as well as for welders, bricklayers, concrete finishers, painters, iron, and insulation workers, electrical and mechanical engineers, technicians, scientists and surveyors.

Fred Dubin, president of an engineering, planning and management firm which has conducted comprehensive energy analyses for many parts of the country, found that two billion dollars invested in energy conservation and solar power provides four times as many jobs as if it were invested in nuclear reactors (64,000 as compared to 15,000).[110] Skip Laitner has shown that about 2½ more jobs are required for solar-developed energy than for the same amount of energy produced by nuclear fission.[111]

The job mix for the various technologies is different. Nuclear energy utilizes fewer tradespeople per professional scientist or technician than does solar energy: for nuclear, the ratio is about 2 to 1; for solar, it is 9 to 1.[112] In addition, a broader array of skills are necessary for building and maintaining solar systems than for building and maintaining nuclear plants. And, as an ERDA report stated:

> Solar systems provide much more room for small business and geographically dispersed businesses and workers than do some of the more complex systems.[113]

The California Public Policy Center's *Jobs From The Sun* report (1978) concluded that providing solar collectors for California structures "wherever practical" would result in 103,000 direct jobs per year between 1981 and 1990. There would be 20,500 jobs in collector manufacture, and 82,300 in installation. Over the ten year period, 1,030,000 job-years of work would be created, along with more than twice as many indirect jobs.

The Jet Propulsion Laboratory in California, in its *Solar Energy in Buildings* report of 1977, compared the jobs which would be created with liquefied natural gas (LNG) use, and by solar energy. In direct jobs, over a twenty year period, the lab concluded there would be 50-80 times more jobs coming from the solar technologies. And, the report discovered, the solar scenario would cost 1.5 times *less* than that utilizing LNG.

The Congressional Office of Technology Assessment (OTA) has examined the job impacts of both coal and solar systems capable of producing equivalent amounts of energy. In its massive report, *Application of Solar Technology to Today's Needs*, published in 1977, OTA documented that an 800 megawatt coal-fired facility would require 2060 total jobs (in man-hours per megawatt year) for construction, operation and maintenance. The same energy from solar hot water collectors would create 3200-4900 total jobs; from photovoltaic systems: 11,200-14,700 jobs.

A report to the New York State Legislative Commission on Energy Systems calculated that the operation and maintenance of large wind systems requires 2-4 times the labor force on a continuous basis as do nuclear fission or coal-fired systems. And the building of wind systems, said the report, would provide employment in generator and electrical component manufacture; sheet metal and structural steel fabrication; cement and wire production. The combined continuous labor requirements for construction and operation of wind systems without storage, according to the report, is 7000 person-years more than for an equivalent nuclear plant. For wind systems with storage, 19,000 more person-years of work are created than are created by an equivalent nuclear plant. The report emphasized:

> The construction phase for both wind with storage and wind without storage would employ thousands of electricians, engineers, heavy equipment operators, laborers, steel workers and other construction personnel.[114]

University of Massachusetts Professor David R. Inglis has determined that if windpower had a share of federal financial support commensurate with its promise, and if the money were spent "to encourage the rapid growth of a wind-turbine building industry," such an effort would utilize industrial facilities which already exist: motor companies which make gears, aircraft companies which make blades, electric companies which make generators.[115] The FEA Project Independence Study estimated that to generate 23% of the nation's electricity with windpower, 140,000 person-years of employment would be created by 1985, and 245,000 by 1990.[116]

The Lockheed Corporation, according to the May, 1978 issue of *RAIN Magazine*, has predicted that when the production of wind energy machines reaches 8,000 per year, the *direct* labor requirement will reach 645,000 jobs. Of these, 89.4% will be in heavy manufacturing, and 10.6% in heavy construction.

The Project Independence Report estimated that even a low level of electrical production by photovoltaic conversion would result in 3,344,000 jobs by 2000.

Jobs would even be created from increased use of wood fuels. The California energy commission has estimated that within three years, 150-200 megawatts of power derived from wood and agricultural wastes would create 3800 jobs for Californians:

1200 in manufacturing; 2400 in construction; and 200 in operation and maintenance.

In an important study issued by the Joint Economic Committee of Congress (*The Employment Impact of the Solar Transition*, Spring, 1979), Dr. Leonard Rodberg argues that

> the same practices which are creating the energy shortage have also been responsible for the shortage of jobs. . . . Solar energy could be the technology that lifts the economy out of the doldrums of the 1970's and into a more prosperous period of the 1980's.

Dr. Rodberg prepared for the Joint Economic Committee a detailed scenario which calls for gradually-increasing investment in conservation and solar development for residential, commercial and industrial use until 1985, followed by a constant level of investment through 1990. A federally-backed energy development bank is the principal financial mechanism for this solar transition which, according to Dr. Rodberg, would create at least 2.9 million direct and indirect jobs. (Even more jobs would be generated through the 'multiplier' effect, that is, the spending of income earned through this new employment, and through the 'accelerator' effect, an increased investment through anticipatory growth, which Rodberg did not quantify.)

According to Rodberg, consumers would be saving tens of billions of dollars by 1990 because they would be using less energy and more renewables, instead of wasting energy and using up fossil fuel. These savings, which greatly exceed the amount needed for solar and conservation investments, would be used to purchase additional goods and services, and to create large numbers of jobs in non-energy related areas.

Rodberg does note that such a major shift in investment towards conservation and renewable energy systems will significantly alter the Department of Labor's current projections regarding where the jobs will be in future energy development. But he asserts that his analysis reveals that *hardly any existing jobs* will actually be lost because of new solar development.

If Dr. Rodberg's analysis is to be of any use, the nation will have to follow his example and begin to plan carefully the smoothest, most equitable transition to alternate energy systems and to the increased energy productivity and the jobs and well-being they will bring.

But today, there does not appear to exist among the nation's official energy planners, among investors, or in the energy industry the interest in expediting and facilitating this transition. Edward Carlough of the Sheet Metal Workers' International Association has reflected the frustration of many regarding the slow pace of solar energy use when he said:

> Dollars now being spent on fossil fuels from abroad would be better spent reducing joblessness by harnessing less expensive, inexhaustible energies at home. It is a national folly that, while the need for energy conservation and solar energy is so great, our sheet metal craftsmen, who can make energy-efficient buildings a reality, sit idle.[117]

Why, with the promise of solar energy so great, has progress been so slow?

The Politics of Solar Energy

In 1952, the Paley Commission noted:

> Efforts made to date to harness solar energy economically are infinitessimal. It's time for aggressive research in the whole field of solar energy — an effort in which the U.S. could make an immense contribution to the welfare of the whole world.[118]

According to the Senate Select Committee on Small Business, the Paley Commission's enthusiastic predictions and their recommendations went "tragically unheeded." And more than 20 years later, new studies for the government done by the General Electric Corporation, by Westinghouse, and by another large energy company — TRW — predicted that the energy contributions of solar technologies by the year 2000 would be minimal: General Electric's figure for solar heating and cooling was 1.6%; Westinghouse's was 3.14%; TRW's was 5.77%.[119]

Senator Gaylord Nelson (D-Wis.), noting that these solar projection studies were performed by companies heavily invested in nuclear power, coal, natural gas and oil, commented:

> The suspicion is unavoidable that these and other absurdly low estimates of the solar contribution during the next 25 years are not of what the estimators think the country *could* do if it put forward anything like the money and effort that went into nuclear development or moonshots, but rather what they hope the country *will* do. Not because doing so little is in the best interests of the great majority of Americans and other people of the world, but because doing more could possibly threaten existing investments in other technologies.[120]

James Piper, President of the Piper Hydro Company, testified before Senator Nelson's Committee on Small Business:

> I do not think Westinghouse can put its heart and soul into producing a good solar system.[121]

Jerry Plunket, a solar businessman from Colorado, noted the difficulties of dealing with the federal government on solar energy matters:

> Rather than assisting us, we have found federal employees unable to understand the solar-state-of-the-technology, unable to formulate

> reasonable plans for moving solar technology ahead, and in fact engaging in projects that were designed to keep university professors employed and off the street, and to use study contracts granted to large firms to make solar energy appear long term, remote and unlikely to respond to our current energy crisis. Further, incredibly expensive solar heating experiments were conducted that seemed designed to explore the upper bounds of costs — not to demonstrate cost effectiveness.[122]

The amount of federal subsidy to stimulate energy production has been enormous. A study from the Battelle Institute reviews several federal incentives to energy production. It shows that over the past 30 years, federal incentives to nuclear power, excluding the costs of federal liability insurance, cost taxpayers between $15 and $17 billion. An estimated $77.2 billion (60% of the total subsidies) went to the oil industry. Natural gas companies received an estimated $15.1 billion, and between 1954 and 1976 the coal industry received $3 billion.

By comparison, the degree of federal support for solar energy production is negligible. Solar energy, however, is expected to compete economically with these other technologies.

The FEA Project Independence Task Force found several "factors which have inhibited" the growth of solar systems. They included: (1) a lack of federal or private industry interest, and (2) underpriced, taxpayer subsidized fuels.[123]

An excellent example of federal indifference to solar energy development occurred recently in Fairfax County, Virginia. A school was completed there which is 100% heated and cooled by solar energy — the nation's first. Fairfax was unable to obtain government assistance for this project, so it sought money for a long time from other sources. Finally, the county received some funds from a university — in Saudi Arabia.[124]

Government officials frequently say that modest sums appropriated for solar energy are as much as can be prudently spent,

Table IV[125]

Year	Number of aircraft produced
1938	3,623
1939	5,856
1940	12,804
1941	26,277
1942	47,836
1943	85,898
1944	96,318

Aircraft production in the U.S.: Production of wind energy systems could soar just as dramatically — *if* the government made it a priority.

given the state of the art. They say that a "new" industry can progress only a little at a time. But on some occasions when this country needed vigorous action, even the federal bureaucracies and private industry responded. For example, when specialized aircraft were needed during World War II, the fledgling aircraft industry was able to increase production from 3,623 in 1938 to 96,318 in 1944. See Table IV.

The FEA concluded:

> It is assumed that wind energy systems are not more difficult and probably far less difficult to produce than World War II aircraft.

No federal funds went to solar research and development until 1970. Since then, the solar budget has been increased each year. See Table V.

The Carter Administration, contrary to its rhetoric, has not significantly increased federal funding for solar energy. For fiscal year 1978, Carter doubled the conservation allocation, increased the solar request slightly, and decreased funds for the

Table V[126]

Year	Allocation
1970	$ .2 million
1971	1.2 million
1972	1.7 million
1973	4.0 million
1974	15.0 million
1975	43.0 million
1976	114.6 million
1977	290.0 million

Federal budget for solar research and development.

Chart III[127]

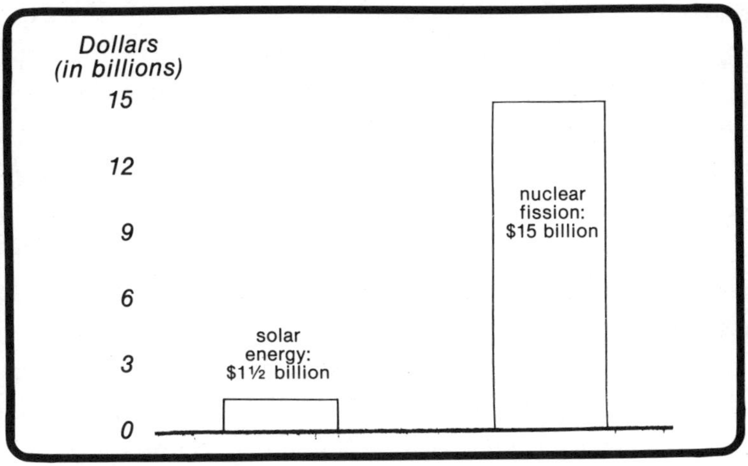

Federal expenditures for solar energy are dwarfed by expenditures for nuclear fission.

Table VI[128]

Budget outlays (in millions)	Technologies
$244	conservation
522	fossil fuels
250	solar
392	fusion
1715	fission

By looking at the proposed FY 1978 budget, it becomes clear that ERDA's priorities were the development of nuclear and fossil fuels — not conservation or renewable solar sources.

liquid metal fast breeder reactor project by about one-third, compared to the Ford Administration's fiscal year 1977 budget. But other categories related to nuclear fission were increased. According to Tom Cochran of the Natural Resources Defense Council, nuclear fission absorbed about 40% of ERDA's budget increases.

If one combines the various categories involving nuclear fission and solar energy, the total 1978 ERDA budget for energy research and development is shown in Table VI.

President Carter's proposed funding of nuclear fission and nuclear fusion technologies thus ended up about *four times* that of conservation and solar technologies combined, despite the significant advantages in economy, employment, safety, and near-term feasibility of conservation and solar energy.

For fiscal year 1980, the Administration's solar budget once again shows some absolute increases, and because previous allocation to solar are so low, they show a very high percentage increase. But the nation is not getting even the billion dollars that the President and Energy Secretary Schlesinger promised: the DOE requested just $750 million, from which the President's Office of Management and Budget promptly cut $100 million.

According to the Solar Lobby, the total 1980 solar request represents less than 8% of the DOE budget. To no one's surprise, atomic weaponry plus research and development for commercial nuclear power are the principal benefactors of DOE's devotion. Weaponry consumes fully 35% of the Department's budget, and commercial nuclear activities another 25%. And, incredibly enough, although Carter has opposed construction of the pilot liquid metal fast breeder reactor, DOE still has requested $504 million for fast breeder research and development during 1980 — almost as much as DOE's entire solar request.

Conclusion

A sadly-neglected ERDA report of a few years ago recognized that among other aspects of the nation's energy dilemma, unwarranted "fear of unemployment is a key political fact."[129] This will be true as long as energy monopolies insist on threatening economic depression and unemployment if their expansion of vast, complex, costly and centralized energy systems is not permitted to continue.

To be sure, jobs will trickle down as a result of investment in wasteful and dangerous energy systems. ERDA has estimated that the current total employment in nuclear fission electric activities is about 80,000 people: mostly engineers, mathematicians, physical and earth scientists, technicians, welders, plus "all other employees."[130] Getty Oil Company's Nuclear Fuel Service Facility, between 1966-1971, employed an average of 1400 temporary workers each year at radioactive "hot spots." To repair six 4½ inch hot water pipes in radioactive areas of a nuclear reactor, Consolidated Edison Company brought 1500 welders, each of whom worked 15 minutes — at which point each had received their maximum permissible dose of radiation. Professor Irwin C. Bupp of MIT has calculated that proposed floating nuclear plants will create jobs beyond those created by land-based nuclear plants — jobs in hull scraping and ferrying.

Promoters of ever-expanding deployment of nuclear and other huge energy systems as the primary means of providing employment and prosperity (such as *Americans for Energy Independence*) try hard to portray those seeking energy efficiency and the commercialization of solar technologies as being against "economic growth . . . workers, the poor, and the disadvantaged."[131]

But it is the energy expansion scenario which wastes both capital and resources. It provides only limited jobs and unreliable energy sources, it causes disease and environmental destruction, and it is not in the interests of "workers, the poor and the disadvantaged," or in the interests of anyone else except the large energy corporations themselves. There are more jobs, and safer jobs, and there will be greater prosperity more evenly distributed, if the nation cuts back significantly on energy waste and moves vigorously toward solar energy. There will also be much less social and political havoc arising from this path to energy sufficiency.

Clearly, those who seek this solution do not seek an era of freezing and starving in the dark. They envision just the opposite: a time of decreasing dependence on foreign countries and on vulnerable and speculative energy systems; a time of abundant jobs and healthier people who live amidst cleaner air and water; and a time when people have greater control over their own lives and more resources with which to obtain the goods and services which make living easier and more enjoyable for all. A transition to a conservation and solar economy, during which no group of people will be made to bear the burdens of changing social values and technological innovations, would mean that the entire society would benefit greatly.

Energy corporation supporters try to suggest that the average American has no business getting involved in energy problems and solutions. *Americans For Energy Independence,* for example, advises citizens to defer to "representatives" in government, labor and industry:

> But [proponents of limited nuclear energy] do not hold positions that would make them accountable for the consequences of limiting the country's economic growth... On the other hand, those representatives of government, labor and industry who have personal responsibility for maintaining the economic well-being of the nation and improving the quality of life for the workers, the poor, and the disadvantaged, are firmly convinced that economic growth should not be limited at all.[132]

As we have demonstrated, there *are* representatives of government, labor and industry who know the great value of decreasing energy waste and maximizing the use of solar technologies, *especially* for the benefit of "the workers, the poor, and the disadvantaged."

But beyond this, the "responsibility" for the economic well-being of the country cannot be the special province of the country's leaders, as Jimmy Carter's budgets make clear. On the contrary, *all Americans* share such responsibility. As the ERDA assessment of solar energy stressed:

> Solar energy is part of a broad societal choice involving much more than the selection of an energy technology.[133]

Indeed, these energy and jobs issues are components of basic struggles over political and economic power. The President of the one-million-member Association of Machinists, William Winpisinger, explained this very well in a speech in Boston in March, 1978:

> It is time to realize that what we are really talking about, in terms of control, is economic and political power. To put it bluntly, the multinational energy corporations have control, have the power, and the government and the people do not. . . . Structural challenge to and changes in private control of energy investment are probably preconditions to any political possibility of solving the energy crisis in an equitable and just manner. [This is] why we must look toward public, as opposed to private, development of alternate energy sources.

Many Americans are ahead of their leaders in understanding the causes of the nation's energy and unemployment problems. They are willing to seek solutions which may not necessarily coincide with corporate myths. They realize that energy efficiency and solar technologies can assure the public of safe and adequate energy supplies; that the people will be able to control its production and use; and that there will be enough jobs available in diverse activities throughout a prosperous nation.

chapter 5 Inaction at the Departments of Energy and Labor

Congressional hearings have made it clear: the Departments of Energy and Labor have made no serious effort to look at the employment consequences of energy policies.

Authors' note: *During our research into energy and employment, we were struck by the federal goverment's lack of commitment — and lack of ability — to coordinate research and planning in these two important fields.*

Since the 1973 Arab oil embargo, Americans have been acutely aware of how serious are our energy and employment problems. Yet, no effort was made by the United States government to discover how both problems could be tackled in concert. In fact, when we first began to examine these issues in 1976, we could not find anyone in the various federal departments with responsibility for either employment or energy planning who had given much thought to the relationship between jobs and energy. Apparently, federal policy makers were content just to go along with the lavishly-advertised claims of the energy and investment communities (not renowned for strong interest either in safe, affordable energy or in full employment) that jobs were to be supplied to Americans only with massive expansion of costly and polluting energy systems.

Never mind that the Employment Act of 1946 committed the United States to policies which ensured "maximum employment"; or that an increasing number of analysts were claiming that existing energy policies were **causing** *unemployment in some parts of the country. Never mind that the buying of foreign oil was generating great inflation here at home; that the nation was paying a huge — if not immediately obvious — price in health, environmental and social damage caused by fossil and nuclear energy development; and that citizens around the country were organizing in opposition to what they regarded as unsafe and economically unsound power stations. The bankers and the power plant contractors and utility companies were telling us they had the answer to the jobs and energy problems. And federal officials — elected representatives and bureaucrats alike — swallowed their pitch without a murmur.*

This is why hearings organized by the Energy Subcommittee of the Joint Economic Committee of Congress, "Creating Jobs Through Energy Policy," were so important. Held on March 15 and 16, 1978, following publication of **Jobs & Energy** *(revised and updated here as the first four chapters of this book) and other studies, the hearings exposed the insensitivity, the resistance, the lack of interest, and the fumbling ineptness of federal officials in making jobs and energy policies.*

In his opening remarks, Subcommittee Chairman Senator Edward Kennedy noted:

> *I could not see where the national energy plan took employment into account. I wondered if the need to expand employment was con-*

> *sidered when the plan was written. I suspect, without knowing, that too little was known about the subject, then and now.*

Extensive testimony at those hearings demonstrated that Senator Kennedy's suspicions were well-founded.

In this chapter, we present an account of those two days of hearings. Following this report, we review what little progress has been made in the energy and employment realm during the year since those seminal hearings of the Joint Economic Committee.

☆ ☆ ☆

"We are appalled to learn there is no definitive analysis or study for the nation on a macro basis [of the impact of energy inputs and alternate energy sources on employment]. The Department of Energy has made no such study. The Office of Management and Budget has made no study. The Commerce Department has no study. The Bureau of Labor Statistics has no study. And the Congressional Research Service has no energy-employment analysis or study."

So declared an angry William Winpisinger, President of the International Association of Machinists and Aerospace Workers (IAM) on March 15, 1978 before the Energy Subcommittee of the Congressional Joint Economic Committee.

Leaders of other unions and of the National Urban League, along with representatives of California Governor Jerry Brown and the City of Hartford, joined Winpisinger in attacking the Administration's lack of integrated energy-employment planning. Confronted with these criticisms and testimony from independent researchers describing numerous non-governmental studies which strongly favored conservation and solar energy development, Secretary of Labor Ray Marshall and high Department of Energy (DOE) officials finally conceded that:

- the Department of Labor (DOL) had essentially no involvement in the formulation of the National Energy Plan (NEP);
- despite the Administration's talk of full employment and energy sufficiency, there is at DOL and DOE very little interest in or capability of devising coordinated plans with which to attain those goals.

With the exception of the NEP's new and important attention to conservation (that is, to increasing the efficiency of energy use), the Plan reflects the energy demand projections and the technological aspirations of corporate energy interests, reinforced by fading dreams of old Atomic Energy Commission employees. In the conservation realm, the Administration has been able to nurture a major theoretical shift away from old orthodoxy by relying on analyses (primarily the 1974 Ford Foundation's *A Time to Choose*) and industrial experience which demonstrate the considerable economic and employment benefits of decreased energy waste. Although neither DOL nor DOE has produced much independent work on the employment impact of energy conservation, energy policy makers in this Administration apparently no longer believe that vigorous conservation would be disastrous for the economy. Even Energy Secretary James Schlesinger has come around on this point (see page 113).

With regard to new energy development, however, government policy makers have been at the analytical mercy of the energy industry. For years, this industry has been advertising extravagantly to successive administrations, to .Congress, union leaders and the public, that its priority scenario of nuclear, synthetic fuel and fossil plant expansion was the *only way* the nation could avoid economic chaos and mass unemployment. The very candid admissions of DOL and DOE officials to this Subcommittee, however, confirmed what Environmentalists for Full Employment and other organizations have been claiming: that although NEP declares it is ''an axiom of public policy that full employment be promoted,'' responsible federal agencies and administration officials have not bothered to verify the industry's claim that certain energy technologies can or cannot fuel the economy, and can or cannot provide the path to full employment.

Yet detailed testimony at these hearings from sources *outside* the government and *outside* the corporate energy industry indicated that a vigorous conservation-plus-solar technology program is indeed able to fuel our economy. Moreover, according to the witnesses, such a program provides many more jobs than the energy industry's scenario. This approach also imposes far less harm to energy industry employees, along with less damage to public health and to the environment, than the nuclear-synfuel-fossil alternative.

Despite talk of new interest in energy-employment relationships, and claims that new studies were either being designed or initiated, DOE and DOL officials at these hearings offered no new data of their own on the potential of solar technologies and on the comparative employment impacts of solar, energy efficiency, coal liquefaction, coal gasification or nuclear technologies. The presentations of these officials were embarrassingly scanty: government experts were no better prepared for these hearings, or for the designing of Part II of the NEP, than they were for the birth of the original NEP last spring.

Senator George McGovern (Dem., SD), on the other hand, quickly went to the heart of the issue:

> I think the energy question is extremely important, but I think the unemployment [problem] is so critical and so painful that if we can relate these two issues together in a way that developing the best possible energy policy would also create the maximum number of good jobs, this is really what we are about.

While Labor Secretary Marshall agreed with McGovern on this point, McGovern went on to note that mention of solar technologies as energy and employment generators were consistently omitted from testimony of Administration witnesses. He then asked Marshall:

> Supposing you were to consider an energy future that was based rather substantially on the development of renewable sources of energy: solar power, wind power, geothermal, conversion of waste materials to energy. What, for example, would be the job-creating impact of an energy design of that kind as over against one that relied primarily on the more conventional sources of energy and possibly heading eventually towards greater reliance on nuclear power? What would be the implications nationally of a nuclear future as over against a solar future in terms of jobs?. . . Aren't these the kinds of studies that the DOL ought to be undertaking in conjunction with the DOE?

Marshall responded that yes, these studies ought to be done, especially since "there is little data on the employment impact of solar and wind."

McGovern also asked Marshall if he thought the Carter Administration had a serious commitment to solar energy, considering the solar budget for the next year had been cut and the Administration's goals were less ambitious than those Governor Jerry Brown had for California. Marshall said that he did believe the Administration was serious about solar technologies. But *neither his testimony, nor that of DOE Assistant Secretary for Policy and Evaluation Alvin Alm,* supports Marshall's claim. While Marshall at least paid some lip service to the need to develop renewable energy sources, Alm did not even mention solar energy as a useful or significant source for the future. Rather, he concentrated on conservation, although even there he offered little that was new: his testimony consisted mostly of warmed-over excerpts of a speech written the previous October by Clark Bullard of the University of Illinois, *prior* to his coming to work at DOE in Alm's office.

When Senator Kennedy asked Alm whether any employment impact analyses of alternate energy sources had been prepared during the formulation of NEP, Alm said: "It is fair to say the evaluation was a very crude calculation." Kennedy then asked whether the Labor Department had contributed to this effort. Alm replied that whatever employment input there had been came from the President's Council of Economic Advisors, *not from DOL.* Alm went on to testify that

> the DOE does not have an adequate capability for analyzing employment impacts of specific conservation investments in a way that permits easy comparison with energy supply investments.

Labor Secretary Marshall was at least forthright about the need for the Administration — and particularly for the Labor Department — to obtain information about the employment impact of energy development:

> I believe that dialogue on national energy issues, for the most part, has omitted serious consideration of the potential employment impact of the proposed NEP... Different energy strategies may yield

> significantly different employment consequences. For this reason, it is important to consider the employment consequences of a particular energy strategy *before* committing ourselves to that strategy... We have seen several dislocations that adversely impact specific areas of our country as we rapidly expand production of coal, gas, oil, uranium or virtually any other energy resource.

Yet when Marshall turned the floor over to his Assistant Secretary for Policy, Evaluation and Research, Arnold Packer, a different Labor Department viewpoint emerged. In contrast to Marshall's candor and responsiveness, Dr. Packer was defensive and obtuse. The problem, he said,

> is, of course, extremely complicated.... There was a general feeling — and I think analysis would bear that out — that any of those alternative approaches, because they were higher [in] cost, would ultimately lead to less employment than [an approach] that was balanced.

Kennedy interrupted to ask if this was Packer's personal view or whether it was based on data which had been submitted to Congress. Packer conceded that he had done no research comparing present energy costs with costs of energy from solar technologies.

Finally, an exasperated Kennedy turned to Marshall and said:

> I don't know whether I have been able to make myself clear... I hope you can understand what we are driving at in terms of how we would like to work with the Department [of Labor].

Marshall indicated that it was clear enough to him, and that the Labor Department shared Kennedy's concerns. Referring to the testimony of James Benson of the Council on Economic Priorities (whose charts showing greater employment on Long Island from solar-plus-conservation technologies than from nuclear plants were displayed prominently in the hearing room), Marshall added: "This kind of analysis is a good beginning."

Benson and the others who comprised a panel of independent analysts were well-prepared, thorough, and open to suggestions and criticisms. Dr. Bruce Hannon of the Center for Advanced Computation at the University of Illinois; Dr. Duane Chapman of Cornell and the Energy Resources Group at the University of California at Berkeley; and Benson, all denounced government agencies for their neglect of energy-employment relationships. Dr. Hannon, one of the pioneers in this field, and the author of seminal employment impact studies concerning transferring highway trust fund money to mass transit and Army Corps of Engineers money to human needs, was especially critical of the resistance he had encountered merely trying to get DOL and DOE officials to examine his work. Hannon pointed out that although traditional wisdom regarding the economic impact of conservation had changed with the coming of the Carter Administration, the federal government "has not really adopted a policy of energy conservation." Perhaps, Hannon added, "such a policy can finally be realized when coupled with the idea of increasing employment as well as conserving energy."

Referring to his analysis of hundreds of specific consumer decisions, and of dollar, energy and employment demands, Hannon maintained that jobs are lost throughout the economy as primary energy is transformed into electricity. He stated that new jobs *have been created* during times when energy growth was in decline. On this point, he was supported by experience in California, according to Wilson Clark, Assistant to Governor Brown for Issues and Planning. Clark testified that there was a 50% decline in California's electricity growth rate during 1977

— and yet personal income in California increased 12.5%, and nearly 500,000 new jobs were created.

Chapman corroborated Hannon's conclusions. He testified that a number of new studies indicate to him that

> policies which accelerate energy use increase unemployment in the long run, and policies which promote growth in conservation, solar energy, renewable resource use and public transportation will increase employment.

Chapman also maintained that the *absence of growth* of solar technologies and renewable resource use had a significant inflationary impact, and that it was necessary for the economy to "adapt to present circumstances by securing growth which reduces energy use and increases employment." Home heating with solar energy, said Chapman, appeared to require at least four times — and probably more than four times — the labor that would be involved in generating electricity using coal. Figures from a California State Employment Development Department study, reported to the Subcommittee by Wilson Clark, indicated that the solar-conservation equivalent to nuclear energy provided about six times as many jobs.

Clark, along with Deputy Director of the Washington Bureau of the National Urban League, Maudine Cooper, strongly supported the incorporation of employment impact studies in energy policy development. Cooper went on to express the Urban League's fear that current energy investment was mis-directed, and that without policy change the nation would be unable to reap the employment benefits of an extensive solar and conservation program:

> Investment in our traditional energy enterprises will continue to place an inordinate drain on investment capital so badly needed for investment in productive capacities which create jobs.

Pointing out that the Secretary of Labor was not even chosen to be a member of the President's new Solar Policy Coordinating Committee (on which sat *every other* Cabinet head plus the directors of the Council of Economic Advisors, the Office of Management and Budget, the Office of Science and Technology, the Environmental Protection Agency, the Council on Environmental Quality and NASA), Cooper expressed the Urban League's doubts whether the Carter Administration really cared very much about combining energy and employment planning.

Cooper addressed the need for energy policy to play a key role in urban revitalization. Nick Carbone (City Councilor in Hartford, Connecticut), addressed the same need. Carbone specifically called for a "massive federal, state and local partnership" to develop with public capital a decentralized solar industry and vigorous conservation capability which is beyond the clutches of utility companies. Carbone said such investments would create many thousands of jobs for those who need them: the urban unemployed. "We must understand," he said, "that cities are inherently conserving." The nation must

> create a system which conserves energy, creates jobs and improves the quality of urban life... A national growth and urban policy must make an effective tool of energy. The policy must support the survival of one of our greatest energy resources — our cities.

Carbone told the Subcommittee that Hartford was not waiting for the Administration to take action, but was going ahead with the establishment of municipal corporations to create jobs through conservation and solar energy development.

California's Clark testified that Jerry Brown was not waiting for Carter either. He said that Brown had identified the development of renewable energy sources and the creation of new jobs in energy as the "major priority of his administration." He pointed out that state conservation policies like consumer appliance energy standards and insulation retrofit standards would create thousands of new jobs. California's goal in water and space heating fueled by the sun is 20% of all residences and commercial buildings by 1985 — *more than is mandated by the NEP for the entire nation.* The manufacture of necessary collectors, plus installation, sales and maintenance of the solar systems, according to Clark, would result in the employment of more than 50,000 people, creating "more jobs than now exist in all of California's gas and electric utility industries."

Criticism of DOL's and DOE's solar and employment policies, plus the work of Hannon, Benson, Clark and the other witnesses, made a strong impression on Subcommittee members. In his remarks opening the second day of hearings, Senator Kennedy said:

> This subcommittee heard testimony which was so challenging that I believe both the Congress and the Administration must respond to it when considering energy, employment and tax policy.

Kennedy added that it was clear to him the Carter Administration had produced its National Energy Plan without considering employment:

> Too little attention has been paid to the potential for creating jobs among our most disadvantaged citizens, particularly minorities,

> through conservation efforts in the cities or alternate energy systems. . . . Because of subsidies to energy and capital, we have consistently disadvantaged labor and reduced the number of jobs available. We have made capital less expensive and labor more expensive, creating unemployment. . . . Because of subsidies to utilities, solar energy is put at a disadvantage. Nonetheless, solar water heating and improved housing construction would create three times more jobs [than nuclear plants] on Long Island, and four times more jobs [than nuclear plants] in California, according to the testimony of our witnesses. . . . Conservation is the best job-creator of all. . . . Over the long run, conservation and alternative energy would place tens of billions of dollars in the hands of consumers, driving the economy through consumer spending instead of trickle down economics which places control at the top.

Officials representing three labor unions clearly agreed with Kennedy: the panel of union witnesses kept up the offensive against the Labor and Energy Departments' lack of coordinated energy-employment policies. Representing President Edward Carlough of the Sheet Metal Workers International Association (SMWIA), analyst Stanley Ruttenberg told the Subcommittee that the SMWIA believed the nation should be able to choose energy technologies which "meet both our energy and employment needs." And, he said,

> . . . it is time the blindfolds were removed. We need only look up to the sky for what must certainly be the most promising of new energy sources — the sun.

Tony Mazzocchi, Vice President of the Oil, Chemical and Atomic Workers, spoke on behalf of his union president A.F. Grospiron. "The nation should be *creating, not destroying,* jobs through energy policy," said Mazzocchi. But, he added, just any jobs, or any energy policies, would not suffice. Speaking with the anguish which comes from representing men and women caught in the nation's most deadly workplace environment, Mazzocchi emphasized that the safeguarding of human health must be a primary criterion of any energy and employment policy:

> Full employment is of little use to a worker sick in a hospital or dead in the graveyard. . . . If an energy source is found to be ultimately unsafe or highly destructive of the natural environment, then that energy source should be abandoned!

The Machinists' Winpisinger, and Ruttenberg, also attacked energy policies which neglect human beings. Currently, said Winpisinger, the Administration's NEP

> utterly fails to take into consideration its impact on the human factor. . . We can develop the most profuse energy production and utility program conceivable, but if it results in widespread displacement of human labor, what good will it do?

And Ruttenberg declared:

> It is time to stop soaking working men and women for the benefit of those who already control a sizable portion of the wealth of our country.

Winpisinger urged the development of renewable energy sources especially for the jobs which would be created for the hard core, for low-skilled and for the semi-skilled unemployed. Such development, he said, would in addition

> open up employment for skilled building trades people, machinists, technicians, engineers, architects and scientists. . . . Probably no other program could be designed to better serve the national interest and the people's interest than achieving full employment through development of alternate energy sources. . . . Development of solar, co-generation, biomass, small hydro- and wind-power with a national manpower policy and full employment program, launched on a full-

> scale basis, would create several million jobs in direct employment, with the ripple effect adding a great many more employment opportunities.

Like Hartford's Carbone, Winpisinger called for direct public investment in and ownership of energy and job development projects. The giant energy companies are already integrated both vertically and horizontally, said Winpisinger. Public ownership would not only help stimulate alternate energy development, but would also provide information on cost and performance and other "proprietary" matters so that consumers and government regulators could better know what was going on in the private sector. A public record would be established against which similar operations in the private sector could be measured (which was what the Tennessee Valley Authority — TVA — was supposed to have been all about). We can, said Winpisinger,

> restore the yardstick principle to the energy industry and gain the benefit of competitive influence.

Winpisinger emphasized that public ownership of solar technologies was especially important since the corporate energy interests were already "casting covetous eyes upon the alternatives, particularly solar. Of all energy sources," he warned the Subcommittee, "solar power must be the people's power."

These hearings were remarkable. Representatives of diverse interest groups effectively challenged energy and employment claims of the corporate energy industry by offering for the record an impressive body of data. They exposed the lack of coordinated planning by President Carter on energy and employment policy. And the highest advisors of the President from the Labor and Energy Departments confessed to past inadequacies and current inabilities, and promised the Subcommittee to do better in the future.

But they *never* will do too much better unless the Carter Administration forces the development of independent analytical capability and integrated policy planning. The President used to talk about full employment and energy sufficiency as his highest priorities. But his administration, like administrations before it, has done so very little to develop an ability to analyze energy and employment problems in cohesive fashion, and to coordinate corrective planning. How many more Congressional hearings and independent studies and public protests will be necessary before the White House aggressively addresses:

• the real potential, if pursued with a genuine commitment by people who actually believe in what they are doing, of solar-plus-conservation technologies to fuel our economy;
• the employment potential of alternate energy technologies;
• the comparative impacts on society of different energy technologies, in terms of equity, health and quality of life;
• strategies for transition from one energy source to another which maximize local decision-making and minimize economic, social and employment disruptions?

The President talks about safe energy and employment, but he and Schlesinger work to speed up the construction and licensing of nuclear plants which are unsafe and which destroy jobs. The Department of Energy has plenty of official advisory panels, but none on jobs or solar energy. The Secretary of Labor is unprepared to deal with the jobs and energy equation, and only gets to be a member of the President's newly-formed Cabinet-level solar policy committee because citizen pressure is applied. The Administration complains that the public is not behind the President on energy and employment policies — but why should the public trust the President when he has no facts of his own? Why should the public agree to ambiguously-phrased calls for sacrifice when people increasingly believe they have to battle the Labor and Energy Departments and the President's highest advisors just to get some reliable information, or even to get them to examine new studies and new experiences which run counter to the claims of the energy industry?

The problem is not simply inadequate White House public relations. The problem is a substantive one: the President has yet to comprehend the close relationship between the nation's energy and employment problems and the need for a unified energy and jobs program which stresses conservation and diverse solar systems. Instead, the President has accepted the word of the energy industry and the financial community regarding what is or is not possible, and has delegated authority to his advisors who pretend they've got all the facts.

He should have been at these hearings.

Afterword

Shortly after the hearings, Senator Kennedy wrote to Secretaries Schlesinger and Marshall requesting them to report how their departments were going to gain necessary expertise, and begin cooperating on energy and employment planning.

Both departments responded with enthusiasm. Deputy Energy Secretary John O'Leary even went so far as to declare:

> [Energy and job planning] is one of our major objectives and we are making an extensive effort in this area. . . We agree that information

> on employment impacts of energy prices and alternate strategies should be available for Congressional consideration. A major effort is now under way to do this.

While O'Leary may have been overstating his zeal, especially with regard to the phrase "major effort," it is clear from the perspective of one year later that the Department of Energy has been more forthcoming in this matter than has the Department of Labor. This is surprising, since at the hearings DOL officials had appeared to be much more cooperative and interested in the issue at hand than their colleagues at DOE.

Unfortunately, aside from allowing an outside consultant to explore DOL's computer model of the U.S. economy at the Bureau of Labor Statistics to determine how it could best be used in energy and employment modelling, DOL has taken *no action.* According to Dr. Willis Nordlund, special assistant to the Under Secretary, "The whole effort just went into a holding pattern." Nordlund maintains that Secretary Marshall and others at DOL continue to believe in the importance of DOL cooperation with DOE, and in both departments' developing expertise in analysis and planning of energy/employment development. But, he said, "severe budget cuts, and cutbacks in staff" just got in the way. "It's really unfortunate. We had grandiose plans. We just don't have the resources."

Nordlund said that the DOL budget for fiscal year 1980 is about $26 billion.

At DOE, officials claim they are now interested in developing ability to do energy and employment analysis, and that they have actually assigned staff responsibility in these areas. We understand that some efforts are now underway to address the deficiencies exposed by the hearings. A number of studies have been initiated, seeking both methodological innovation and quantitative results. Preliminary findings, from the Battelle and Argonne Laboratories, and from the Solar Energy Research Institute, should be forthcoming.

Apparently, however, knowledge of joint energy-employment analysis and planning was so scanty that DOE officials believe they need to concentrate at first on identifying precisely what kinds of information must be obtained, and on developing long-term strategies for dealing with these issues in tandem. Nonetheless, according to DOE consultant Meg Schacter who has been spending much of her time on energy and employment analysis,

> I think DOE is now undertaking an admirable effort to set out the issue on a quantitative level, to analyze the data, and to move on to examining the qualitative aspects.

She, for one, is convinced that qualified people — if only a handful so far — are doing what they can today within DOE's vast and convoluted bureaucracy.

Still, it is one year since Secretary Marshall and Secretary Schlesinger promised to bring their departments together, and the interagency task force they spoke of creating is yet to become a reality. A draft of the required Memorandum of Understanding (MOU) languished at DOE until January 1979, and now languishes at DOL. This MOU, when and if it is approved by both departments, will establish two working groups: one on labor-employment analysis, and one on training. They will be chaired by DOE Under Secretary O'Leary (not known as one of those at DOE with an abiding interest in this matter) and by DOL Under Secretary Robert Brown.

And still, despite the data coming out of the public sector

and labor unions which continue to show enormous benefits — employment and other — for the nation from a solar-plus-conservation economy, officials at DOE and DOL are yet to speak out against the continuing propaganda of the energy industry. As the JEC hearings made clear, the "we have no other choice" line of industry has proved to be an effective strategy, if only because independent analyses have not been done, and because sufficient efforts in alternate energy development and funding have not been made.

One bit of progress which could help change this situation for the better is that shortly after the hearings concluded, Senators Kennedy and McGovern helped to guide legislation into law which requires the Secretary of Labor to prepare job impact statements on proposed energy and conservation laws presented to Congress. This was embodied in Public Law 92-524, the Comprehensive Employment Training Act Amendments of 1978. Unfortunately, precisely how the Secretary will fulfill the requirements of this law in the absence of personnel, funds or technical ability, is unclear.

That the Department of Energy may be moving along somewhat on energy and employment planning is encouraging. But it is not enough. There is little specific knowledge *outside* DOE about what it is doing. The Department has sought virtually no cooperation with the many citizen and public interest and labor groups which have been involved with this issue for several years. Considering the magnitude of the budgets and staffs of both departments (DOE's budget for 1979 is almost $11 billion), and considering what they are capable of doing, it is clear that integrating energy and employment planning has not yet captured the imagination of the nation's highest policy makers.

There remains urgent need to confront the job blackmail of the energy industry, and for the industry, the government and citizens to consider carefully how best to plan the national transition to a solar and conservation economy. Adequate funds and resources must be made available, and extensive work must be done to plan out and effectuate this major shift in national policy.

Richard Grossman and Gail Daneker
Washington DC, March 1979

appendix

New Perspectives on Jobs and Energy

Several short essays about employment, energy use and conservation, and the questionable value of orthodox wisdom.

For goodness' sake why?

If expansion of large, complex energy systems is so wasteful, causes pollution and disease, gobbles up investment dollars, threatens the democratic process, displaces existing jobs and chokes out large numbers of new jobs, why have utilities and energy industries been so determined to expand, expand, expand?

For regulated utilities, there is a clear, direct relationship between energy growth and profits — *their profits.* A utility's rate of return on investment is set by law at a fixed percentage. The more it invests, the more profits it reaps. To attract more investors who will put up the money for more facilities which will return more profits in order to attract more investors, utilities seek to invest in more facilities which will attract more investors and return more profits.

If consumers use decreasing amounts of energy, utilities have a hard time justifying investment in more energy-producing facilities. Less utility money is available for reinvestment, and utilities begin to experience difficulty attracting money from the financial community. Because energy use has decreased (or in some places has increased at a lower rate) since the Arab oil embargo, many utilities have lost large amounts of revenues, and have ended up with unneeded facilities which were planned a decade ago. They have had to seek huge rate increases to pay construction and operation costs of their expensive overbuilding. Thus do consumers rightly complain that it does not make much difference whether or not they turn their thermostats down: they may use less energy, but their bills keep going up anyway.

It is just not in the interests of the utilities to promote energy efficiency very seriously.

Utilities pay some lip service to conservation — especially when there are shortages of fuel or they are having particular problems delivering the energy they are required to deliver. But they do not want to encourage *too much* conservation. A recent Mobil advertisement, for example, had some nice words about conservation, but it strongly stressed the need not to "trade-off" conservation for jobs or GNP growth, and the need to preserve consumers' "freedom of choice" to use energy as they see fit.

Thus, in order to raise their rates to expand and increase profits to expand and raise rates, it has been in the interest of the utilities to have people believe that energy growth leads to employment growth and economic prosperity; to persuade large industries to use energy extravagantly by charging the lowest rates to the largest users; to threaten that without nuclear and other exotic energy systems Americans would soon freeze and starve in the dark; and to oppose environmental and public health measures, which interfere with their visions of unlimited expansion without regulation. In addition, utilities have recently become interested in the plight of the poor and disadvantaged: utilities take great advertising pride in having brought them to the threshold of the good life, and are now declaring it is for the sake of these people that more and more giant, dangerous energy facilities should be built.

Some of the nation's largest manufacturers build the equipment used by the utilities. They also build and sell the appliances that people buy, and the machines that industries buy, all of which consume large amounts of energy. Many of the corporations which control energy resources (coal, oil, gas, uranium) are also in the business of finding, obtaining, transforming and delivering these resources in usable form. The profits of these companies increase as the utilities expand their ability to supply energy. Clearly, then, it is in the interest of the energy-production companies and the energy-equipment manufacturers to throw in their lot with the utilities.

The utility industry, with the full support of energy industries and the government (which has provided research, development and demonstration money, along with favorable subsidies and tax incentives), has been very successful at expanding without much caring about the undesirable societal, economic and human consequences of their efforts. This confirms what John Maynard Keynes noted earlier in the century: *There is no evidence from experience that the investment policy which is socially advantageous coincides with that which is most profitable.*

GNP

The selection of Gross National Product (GNP) as a main criterion for evaluating the nation's economic health has served to mislead the public on energy and employment issues.

GNP is the market value of all goods and services produced in the economy over the course of a year. As far as GNP is concerned, *everything that costs money* is considered a benefit. GNP includes expenditures for desirable items — such as for energy, housing, education, food, etc., — but without taking into account whether they are made available efficiently or safely. GNP also includes costs of items not generally considered to be beneficial to the nation: disease treatment, pollution clean-up, weapons production and sales, wars, as well as unemployment insurance, worker's compensation, welfare payments, etc.

Some analysts believe that the only part of the GNP which is actually increasing these days is that part caused by the costs of pollution, environmental degradation and human suffering caused by wasteful, inefficient and dangerous methods of production (especially of energy).

A recent Harris poll has indicated that a growing number of people believe the quality of life has generally deteriorated over the past decade. But the GNP has been increasing, despite some temporary decreases in "growth" rates during the '74-'75 recession. Thus, although individuals believe the quality of life has gotten worse, according to the GNP, the economy has gotten bigger and better.

For the United States to be striving for ever larger and larger GNP, to be basing energy, employment and other strategies on a deceptive criterion which conceals what is actually taking place in terms of unemployment, energy waste, work-related illness and public morbidity, does not appear to be the height of wisdom.

On the poor and disadvantaged

Energy industries say they must expand for the sake of the poor and disadvantaged. If energy industries were truly concerned about these people, they would be spending money to develop inexpensive, energy-efficient and long-lasting appliances. Yet, as Professor Herman Daly has pointed out, the industry spends *eight times* as much money on advertising as it does on research and development of energy-efficient consumer goods.

In addition, utilities would not be opposing affordable 'life-line' utility rates, and would not be charging higher rates to the smaller users. They would not be opposing clean air and clean water regulations, since foul air and polluted water affect the poor more than those who can afford to escape to cleaner realms. They would not be fighting commercialization of safer, cleaner, cheaper solar energy systems and energy-efficiency techniques (which they usually fight with rate-payer money). They would not constantly be playing down the danger aspects of complex technologies like nuclear fission, coal conversion, the liquefaction and transportation of natural gas, which threaten the poor first and most. And they would not be misleading about where the jobs are.

On experts, truth & profit centers

In just a few years, orthodox wisdom concerning the economic impacts of energy conservation has been turned on its head. What was considered incontrovertible in 1974 is now thought to be as old fashioned as the medical practice of bleeding out the evil humours.

In its definitive 1974 publication entitled *Energy Use to the Year 1983* the Chase Manhattan Bank advised its brethren in the financial community that economic ruin would surely follow attempts to cut back on energy use:

> . . . an analysis of the uses of energy reveals little scope for major reductions without harm to the nation's economy and its standard of living. The great bulk of this energy is utilized for essential purposes — as much as two-thirds for business-related reasons. And much of the remaining third serves essential private needs. Conceivably, the use of energy for such recreational purposes as vacation travel and the viewing of television might be reduced — but not without widespread economic and political repercussions. There are some minor uses of energy that could be regarded as strictly non-essential, but their elimination would not permit any significant change.[134]

ERDA Administrator Robert Seamans subscribed to this point of view in those dark ages of a few years ago. He used to point out that each worker in the United States consumed a certain number of barrels of oil performing a day's work. He therefore concluded that for every decrease in oil imports without equivalent expansion of energy supply, people would be thrown out of work.

The Ford Foundation study on energy and the economy was published in 1974. This report suggested that the nation's energy growth rate could be cut in half without adverse impact upon the economy. Not surprisingly, such a suggestion received little support from the business and financial communities, or from economists and energy experts within the government. But despite its heretical conclusions, this study has gradually become accepted by experts as subsequent analyses and practical experience have provided confirming data. According to a recent article by Anthony Parisi in the New York *Times*, energy economists now believe that

> a fundamental change may be taking place in the way the economy depends on energy... in the future, the nation may achieve healthy economic growth with a much slower growth in energy consumption.[135]

The evidence must be persuasive indeed, for even Energy Secretary Jim Schlesinger has revealed that he himself has embraced the new wisdom:

> If we begin to moderate our energy growth rate now, and decouple the growth in energy consumption from growth in Gross National Product, the United States and the industrialized West and Japan may be able to move through much of the 1980's with minimal long-term disruptions... growth rates may slow somewhat, but they should remain healthy and reasonably predictable.[136]

Today, experts are seeking to explain this remarkable transformation of yesterday's best wisdom. Dr. Alvin Cook, energy analyst at Data Resources, Inc., theorizes that the experts were so very resistant to new thinking on energy efficiency potentials because so little analysis had actually been done. And, speculates Cook, the analyses had not been done because virtually no institutions or businesses believed they would profit from discovering that energy could be used more efficiently without impacting economic growth. Cook noted that commercial interests "create the need for economic studies," and conservation just did not have the "profit center" to encourage that attention be focused to conservation possibilities. As Cook concluded:

> We have oil companies, coal companies, electric companies; we don't have conservation companies.[137]

What neither Cook nor the *Times* suggests is that a "profit center" *did* exist: namely, the American people. The American people were (and still are) fully prepared to accept the financial, employment and health benefits which would result from increased energy conservation. It seems to us that if commercial interests were not investigating conservation potentials (for whatever reasons), then that task should have been taken up on behalf of the American people by Seaman's ERDA and the Department of Labor and the Department of Commerce and the Council of Economic Advisors and the Office of Management and Budget.

The parallel between wisdoms regarding conservation and the wisdoms on the potential of diverse solar energy technologies today is striking. The equivalent to the Chase Manhattan Bank study consists of analyses done by General Electric and Westinghouse for the government in the early 1970's which indicated that by the year 2000, the nation could be deriving from solar heating and cooling 1.6% (GE) and 3.14% (W) of its energy needs. By Cook's logic, these figures are absurdly low because no solar profit center existed in corporate America to see to it that the projections were more realistic.

And yet there *is* a significant solar energy profit center. Unfortunately, it consists "only" of the same American public which did not constitute a proper profit center for energy conservation. But *this time around*, the interests of the American people must be acknowledged. It is therefore essential that the Labor, Energy, Commerce, Agriculture and HUD Departments provide the pioneering studies and affirmations of what can be done with solar technologies. As the Joint Economic Committee hearings showed, there are already large numbers of studies being produced by independent researchers and public interest groups and labor unions which can be used as models for official investigations.

We believe these studies should be done from a positive standpoint. For example, investigators should assume a high percentage of solar capacity for the year 2000 (say, 40-50%), and then seek to determine how this can be achieved. Analysts will have to examine employment, capital, and resource components, along with the need to alter habits, institutions and old energy systems. Assuming such analyses are done by people who believe in what they are doing and who have the resources of the government at their disposal, they should pave the way for innumerable other studies and acceleration of solar construction. But most importantly, they would change the orthodox wisdom of today, and help to open the minds of the "experts" to new data and new possibilities, and new investments, just as the Ford Foundation study paved the way for the emergence of the new wisdom — and new investments — in conservation.

It took only four years for the experts to go from the best

Chase Manhattan had to offer to the current expert view as summarized by Harvard's Dr. Harvey Brooks: "Improved energy efficiency is the key to conservation coupled with economic growth." Perhaps if DOL and DOE and the rest of the government can admit that solar energy is worth developing as long as the general public will benefit, then before long experts in the financial community and in energy realms will be working out the details of an extensive national solar energy transaction. And people will be looking back at the GE and Westinghouse solar predictions with the same bemused nostalgia with which they now view Robert Seamans and the Chase Manhattan Bank.

Footnotes

1. The Congressional Budget Office, *The Disappointing Recovery,* Washington, D.C.: January 1, 1977.
2. Hubert Humphrey, *Congressional Record,* January 10, 1977.
3. Cass Alvin, Energy Conference, Los Angeles County Federation of Labor, January 15, 1977.
4. R. Denny Scott, *The Energy Dilemma — What It Means to Jobs,* Woodworkers' International Union, August 16, 1976.
5. Darrell Huff, *How to Lie With Statistics,* New York: W.W. Norton & Co., 1954.
6. Herman Daly, "Electric Power, Employment and Economic Growth," in *Towards a Steady State Economy,* San Francisco: W.H. Freeman & Company, 1972.
7. A.H. Raskin, *The New York Times,* January 26, 1977.
8. The Ford Foundation, *A Time to Choose,* Cambridge: Ballinger Publishing Company, 1974.
9. *Dollars and Sense,* September 1976.
10. Natural Resources Defense Council, *Choosing an Electrical Energy Future for the Pacific Northwest: An Alternative Scenario,* Palo Alto, 1977.
11. Denis Hayes, *Energy: The Case for Conservation,* Washington, D.C.: Worldwatch Institute, 1976.
12. Ford Foundation, op. cit.
13. Edison Electric Institute, 1973, cited in Schipper, *Energy Conservation: Its Nature, Hidden Benefits and Hidden Barriers,* Berkeley: Lawrence Berkeley Laboratory & the University of California at Berkeley, June 1, 1975.
14. Bruce Hannon, cited in Schipper, op. cit.
15. Ford Foundation, op. cit.
16. Herman Daly, op. cit.
17. *Notes from the Joint Economic Committee,* United States Congress, Washington, D.C.: Volume 3, No. 1, January 21, 1977.
18. *The New York Times,* September 29, 1976.
19. Federal Energy Administration, *Conservation Paper 15,* Washington, D.C.: 1976.
20. Herman Daly, op. cit.
21. Resources for the Future *Newsletter,* Washington, D.C.: Fall 1976, Number 53.
22. Kenneth Watt, et. al., Testimony before the Subcommittee on Energy and Power, U.S. House of Representatives, March 25, 1976.
23. Ford Foundation, op. cit.
24. Cooper and Weinberg, Institute For Energy Analysis, cited in *A Preliminary Social and Environmental Assessment of the ERDA Solar Energy Program 1975-2020.*
25. *Science* magazine, January 27, 1977.
26. United States Energy Resources and Development Administration (ERDA), *A Preliminary Social and Environmental Assessment of the ERDA Solar Energy Program, 1975-2020,* Draft Final Report, July, 1976 (prepared by Stanford Research Institute).
27. *The Conference Board Record,* Volume 12, No. 2, February, 1975.
28. Thomas V. Long & Lee Schipper, Testimony before the Congressional Joint Economic Committee, November 16, 1976.
29. Hayes, Denis, op. cit.
30. Critical Mass: *Jobs and Conservation,* Washington, D.C.: January, 1977.

31. Amory Lovins, testimony before a joint hearing of the Select Committee on Small Business and the Committee on Interior and Insular Affairs, United States Senate, December 9, 1976.

32. Bruce Hannon, *Energy and Labor in the Conserver Society,* July, 1976.

33. ERDA 76-41: *New Technologies for Electric Utilities,* Washington, D.C.: December, 1976; draft final report.

34. Amory Lovins, op. cit.

35. Ibid.

36. Private communication from Edison Electric Institute economist Sue Lerner.

37. Friends of the Earth: *ECO,* San Francisco: October, 1975.

38. The Conference Board: *Capital Invested,* Road Maps of Industry No. 1799, New York: January, 1977.

39. American Institute of Architects: *A Nation of Energy Efficient Buildings By 1990,* New York: 1972.

40. John Holdren, *The New York Times,* July 23, 1975.

41. Bonneville Power Administration, *Electric Energy Conservation Study,* Portland: 1976 (prepared by Skidmore, Owings & Merrill).

42. Ron Lanoue, *Nuclear Plants: The More They Build, The More You Pay,* Washington, DC.: Center for Study of Responsive Law, 1976.

43. Denis Hayes, op. cit.

44. American Institute of Architects, op. cit.

45. Bonneville Power Administration, op. cit.

46. *Illinois Energy Issues,* January, 1977.

47. Federal Energy Administration, *Energy Reporter,* Washington, D.C.: February, 1976.

48. City of Seattle, Department of Lighting, *Seattle 1990,* February, 1976.

49. Dow Chemical Company, *Industrial Energy Study,* a report to the National Science Foundation, June, 1975.

50. Natural Resources Defense Council, *California Energy Policy Project: A Proposal,* Palo Alto: February, 1977.

51. *Sheet Metal Workers' Journal,* February, 1977.

52. Energy Policy Office, State of Massachusetts, *The Use of Solar Energy for Space Heating and Hot Water Systems,* Boston: April, 1976.

53. The State of Colorado, *Energy Conservation Plan,* Denver: 1976.

54. Frank Zarb, testimony before the Congressional Joint Economic Committee, February 3, 1976.

55. Americans For Energy Independence, *Energy Conservation Strategies,* Virginia: 1976.

56. Denis Hayes, op. cit.

57. Bonneville Power Administration, op. cit.

58. Amory Lovins, op. cit.

59. Federal Energy Administration, *Conservation Investment as a Gas Utility Supply Option,* Washington, D.C.: 1976.

60. Critical Mass, *Jobs and Conservation,* Washington, D.C.: January, 1976.

61. American Institute of Architects, op. cit.

62. *Sheet Metal Workers' Journal,* op. cit.

63. Senate Commerce Committee Staff, *An Energy Conservation Public Works Program,* Washington, D.C.: December, 1976.

64. Skip Laitner/Critical Mass, *The Impact of Solar Energy and Conservation Technologies on Employment,* Washington, D.C.: May, 1976.

65. Americans For Energy Independence, op. cit.

66. Federal Energy Administration, *Energy and Economic Impacts of Mandatory Deposits,* Research Triangle Park: September, 1976.

67. Washington Center for Metropolitan Studies, Conference Proceedings: *Capturing the Sun Through Bioconversion,* Washington, D.C.: 1976.

68. *Congressional Record,* October 26, 1976.

69. *San Francisco Examiner,* December 15, 1976.

70. ERDA, *A Preliminary Social and Environmental Assessment of the ERDA Social Energy Program 1975-2020.*

71. Federal Energy Administration, *Project Independence Task Force Report,* Washington, D.C.: 1974.

72. *Energies,* Volume 2, No. 9, Torrance, California: Septemer, 1976.

73. News Release, Sheet Metal Workers' International Association, New York: March 8, 1977.

74. Senate Select Committee on Small Business: Solar Energy Hearings, part II (Ninety-fourth Congress).

75. Ibid.

76. Ibid.

77. *Congressional Record, September 21, 1976.*

78. *Senate Select Committee on Small Business: Solar Energy Hearings, part II.*

79. *Federal Energy Administration, Solar Collector Manufacturing Activity, January-June 1976,* Washington, D.C.

80. Senate Select Committee on Small Business: Solar Energy Hearings, part I.

81. *Congressional Record,* September 21, 1976.

82. State of Oregon, Office of Energy, Research & Planning, *Transition,* Olympia: January, 1975.

83. ERDA, *A Preliminary Social and Environmental Assessment....*

84. Federal Energy Administration, *Project Independence Task Force Report on Solar Energy,* Washington, D.C.: 1974.

85. "Exxon Energy Course for Women—Part 2" — in women's magazines, *passim.*

86. *Congressional Record,* October 26, 1976.

87. News Release, Sheet Metal Workers' International Association, op. cit.

88. Private communication from Roy Beckman, TDC, March 7, 1977.

89. ERDA, *Solar Energy Report,* Washington, D.C.: February 3, 1977.

90. National Science Foundation—RANN, *Wind Machines,* Washington, D.C.: 1976.

91. *Congressional Record,* September 21, 1976.

92. Federal Energy Administration, *Project Independence Task Force on Solar Energy,* Washington, D.C.: 1974.

93. Private communication from Roy Beckman, TDC, March 7, 1977.

94. New York State Energy, Research and Development Authority, News Release, Albany: March, 1977.

95. Washington Center for Metropolitan Studies, op. cit.

96. Ibid.

97. *Congressional Record,* October 26, 1976.

98. Edward J. Carlough, "A Note of Explanation," Sheet Metal Workers' International Association News Release, March 8, 1977.

99. Energy Policy Office, State of Massachusetts, op. cit.

100. Robert Mills, February 15, 1977.

101. Stanford Research Institute, *Strategic Implication of Solar Energy for Employment of Sheet Metal Workers,* Palo Alto: June, 1975.

102. *The Laborer,* "Solar Energy: Source of Power, Key to Jobs," June, 1976.

103. *The Machinist,* "Solar Energy Can Generate A Lot of Heat, Power, Jobs," November, 1976.

104. Ibid.

105. Ibid.

106. R. Denny Scott, op. cit.

107. U.S. Department of Labor, *Worklife,* August, 1976.

108. Federal Energy Administration, *Project Independence Labor Task Force Report,* Washington, D.C.: 1974.

109. Institute for Local Self-Reliance, *Self-Reliance,* Washington, D.C.: September 24, 1976.

110. Private communication, September 24, 1976.

111. Skip Laitner/Critical Mass, op. cit.

112. Federal Energy Administration, *Project Independence Solar Energy Task Force,* Washington, D.C.: 1974.

113. ERDA, *A Preliminary Social and Environmental Assessment...*

114. Monroe, Jackson, et. al., *Energy and Employment in New York State,* A report to the New York State Legislative Committee on Energy Systems, Albany: May 3, 1975 [draft].

115. David R. Inglis, "Wind Power Now," *Bulletin of the Atomic Scientists,* October, 1975.

116. Federal Energy Administration, *Project Independence Report.*

117. Sheet Metal Workers' International Association, News Release, op. cit.

118. Senate Select Committee on Small Business: Solar Energy Hearings.

119. Ibid.

120. Ibid.

121. Ibid.

122. Ibid.

123. Federal Energy Administration, *Project Independence Report.*

124. *The Washington Post,* January 3, 1977.

125. Federal Energy Administration, *Project Independence Report.*

126. Senate Select Committee on Small Business, Solar Energy Hearings.

127. Figures from the Senate Select Committee on Small Business, Solar Energy Hearings, , revised using data from the Battelle Institute (1978) and from Herb Epstein, Center for Renewable Resources, Washington, D.C. (1979).

128. Cochran, Tom, *An Analysis of the Carter Administration FY 1978 Budget,* Natural Resources Defense Council, Washington, D.C.: February 25, 1977.

129. ERDA, *A Preliminary Social and Environmental Assessment...*

130. ERDA, 76-111, *Employment in Nuclear Energy Activities, 1975, A Highlights Report,* Washington, D.C.: August, 1976.

131. Americans for Energy Independence, *Jobs and Energy,* Energy Issues #4, 1976.

132. Ibid.

133. ERDA, *A Preliminary Social and Environmental Assessment...*

134. Hayes, Denis, op. cit.

135. Parisi, Anthony, The New York *Times,* January 8, 1978.

136. Schlesinger, James, testimony before the JEC, March 21, 1978.

137. Parisi, op. cit.

Index

Does your organization have a book it wants to publish?

Alyson Publications, the publisher of this book, is affiliated with Carrier Pigeon, a distributor of progressive and feminist books and magazines. Both as a publisher and as a distributor our basic goal is the same: to help well-written literature about important social issues reach a wider audience.

Besides publishing books in the traditional sense, we also do joint book publishing with interested organizations and individuals. If you have a book you want to publish, but lack the skills, resources, or sales network to publish it alone, we may be able to help.

In a typical co-publishing arrangement, we would be responsible for most of the design, production and publishing work, but the organization or person behind the book would share some of the production costs and would get a comparable part of the print run. Thus, you get copies of your book at a far lower unit cost than if you published it yourselves, and in addition you've got our distribution network complementing your efforts. This cooperative arrangement makes it possible to publish many important books that would never be commercially feasible otherwise.

The exact arrangements will depend on the contents of the book, its length, and your needs. For more information, please write to Alyson Publications, Inc., 75 Kneeland St., Room 309, Dept. 362, Boston, Mass. 02111; or phone (617) 542-5679.

The authors of this book are on the staff of Environmentalists for Full Employment, a coalition of several hundred environmental and community action groups working for both a full employment economy and a clean, healthy society.

EFFE was formed three years ago to help provide information on the employment impact of public health and environmental protection policies, and to build bridges between sectors of our society that are at large with one another over these issues.

The evidence is strong that with well-planned transitions, there need be *no* conflict between socially useful employment, healthy workplaces and clean, natural environments. But these transitions must include preservation of job equity, concern for job dislocations, and a wide range of effective employment and re-employment assistance guarantees.

EFFE believes that environmentalists as one interest group cannot force the changes this country requires. Neither, working alone, can organized or unorganized labor, minorities or the poor. But should these groups join to bring about social change —in such a way that no one sector unfairly bears the burdens, and no one group unfairly benefits—then we will find there can be increased opportunities for the good life for all.

If you'd like more information about Environmentalists for Full Employment, write to them at: 1536 16th St. NW, 1st floor, Washington, D.C. 20036.

For more copies of this book, see the next page.

For more copies of this book:

Energy, Jobs and the Economy is valuable reading for any groups that are affected by or concerned about unemployment and energy usage. Unions, students, and workers in energy-related fields will find it especially useful.

The discount rates below apply to all bulk orders. Please include payment with your order. These prices are all postpaid; add $1 for rush orders.

2 to 4 copies: $3.10 each
5 to 10 copies: $2.80 each
11 to 20 copies: $2.30 each
21 to 50 copies: $2.00 each
51 to 200 copies: $1.80 each

For larger quantities, please write.

Send orders to Carrier Pigeon
75 Kneeland St., room 309
Boston, Mass. 02111